Aquatic Environmental Chemistry

Alan G. Howard

Senior Lecturer, Department of Chemistry,
University of Southampton

Series sponsor: **ZENECA**

ZENECA is a major international company active in four main areas of business: Pharmaceuticals, Agrochemicals and Seeds, Specialty Chemicals, and Biological Products.

ZENECA's skill and innovative ideas in organic chemistry and bioscience create products and services which improve the world's health, nutrition, environment, and quality of life.

ZENECA is committed to the support of education in chemistry and chemical engineering.

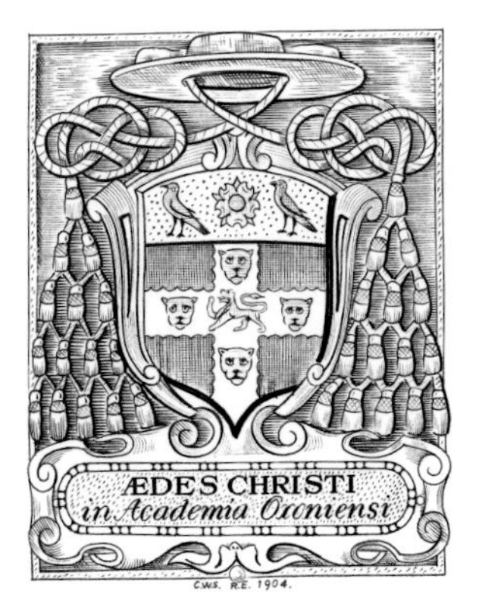

OXFORD NEW YORK TOKYO
OXFORD UNIVERSITY PRESS
1998

Oxford University Press, Great Clarendon Street, Oxford OX2 6DP

Oxford New York
Athens Auckland Bangkok Bogota Bombay
Buenos Aires Calcutta Cape Town Dar es Salaam
Delhi Florence Hong Kong Istanbul Karachi
Kuala Lumpur Madras Madrid Melbourne
Mexico City Nairobi Paris Singapore
Taipei Tokyo Toronto Warsaw

and associated companies in
Berlin Ibadan

Oxford is a trade mark of Oxford University Press

Published in the United States
by Oxford University Press Inc., New York

A catalogue record for this book is available from the British Library

Library of Congress Cataloging in Publication Data

Howard, A. G. (Alan G.)
Aquatic environmental chemistry / Alan G. Howard.
(Oxford chemistry primers; 57)
Includes bibliographical references and index.
1. Water chemistry. I. Title. II. Series.
GB855.H68 1998 628.1'68–dc21 97-42913
ISBN 0 19 850283 4 (Pbk)

Typeset by EXPO Holdings, Malaysia

Printed in Great Britain by
The Bath Press, Avon

Series Editor's Foreword

The state of natural waters is a subject of vital concern to all of us. Establishing the effects of humanity's activities on them can only follow an understanding of their intrinsic properties. Many of these can be explained using the bases of inorganic aqueous solution chemistry and it is the theme of this book to provide these principles and to illustrate their application.

Oxford Chemistry Primers are designed to give a concise introduction to all chemistry students by providing the material that would normally be covered in an 8–10-lecture course. As well as giving up-to-date information, this series provides explanations and rationales that form the framework of an understanding of inorganic chemistry. Alan Howard here gives us the basis of the course he has successfully introduced in Southampton over the last three years. He shows how the fundamentals of equilibrium inorganic chemistry can be used to account for the properties of natural waters in their varied environments.

John Evans

Department of Chemistry, University of Southampton

Preface

Environmental chemistry courses have become an increasingly popular option in degree programmes throughout the world. The reasons for their introduction, whilst diverse, are underpinned by a growing awareness of environmental problems and a desire to understand the behaviour of natural and polluted systems. From the viewpoint of an educator, the environmental link can sometimes provide a valuable boost to a student who can gain inspiration from the reality of the subject.

This book has been written to provide an introduction to equilibrium inorganic chemistry by drawing upon examples from the aquatic environment. It is designed for readers having the levels of basic chemical and mathematical ability generally expected of university entrants. In most UK degree structures it will be seen to cover first year material for students studying chemistry, environmental science, geology, irrigation science, and oceanography.

This is a basic introductory text which attempts to cover the underlying properties of both freshwater and marine systems without placing a particular emphasis on either. Two aspects have been deliberately compromised because of the size and level of the book. First, this is a text on equilibrium inorganic chemistry. Whilst it is recognized that a large number of aquatic systems are not at equilibrium, the level of confusion that can be induced by the simultaneous presentation of equilibrium and kinetic treatments would not be justifiable for the target readership. The second compromise arises from whether the concept of activities should be covered in any detail. In general the book employs concentrations, and not activities, but where the concentration/activity problem arises it is commented upon.

The format adopted for the book consists of two parallel columns. The inner column is the main body of the book and the reader should not need to use the outer column at all to cover the main material. The outer column provides secondary material such as reminders or comments and is sometimes employed to lead the reader through what, to some, might be a tricky derivation.

Southampton A. G. H.
December 1997

Acknowledgements

The author would like to thank all those who so freely gave of their time to read the initial 'drafty' manuscript. In particular I would like to thank John Evans, Sandra Dann, Gillian Reid, and Peter Statham for their detailed comments which have led to significant improvements being made to the manuscript. My thanks are also due to Roger Parsons for his help in developing an approach to the problem, unsolvable at this level, of whether to assign units to equilibrium constants.

Southampton A. G. H.
September 1997

Contents

1 The Earth and biogeochemical cycling

1.1 The structure of the Earth

The fractionation of material during and after the formation of the Earth has given it a layered structure. Whilst it is relatively straightforward for us to investigate the nature of the Earth's surface, gaining access to its deeper layers is at present impossible. Our current understanding of the composition of material at the centre of the Earth is therefore restricted to that which can be inferred from indirect physical measurements. The outermost layer of the solid Earth, its crust (Fig. 1.1), is the most familiar to man and is the part of the Earth most intimately associated with the aquatic environment.

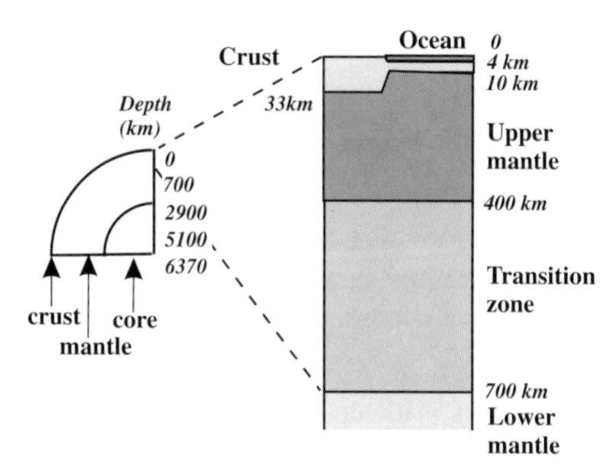

Fig. 1.1 The structure of the Earth.

The boundary between the crust and the mantle was first identified by the discovery of a sudden increase in the velocity of seismic waves at the boundary. This discovery was made by Andrija Mohorovičic and the boundary is now named after him—the Mohorovičic discontinuity or Moho.

The crust is relatively thin. Beneath the oceans it is at its thinnest (5–10 km) but it increases to between 30 km under the flat plains, reaching *c.* 65 km under some mountains. Chemically, the crust is made up from a variety of rock types, many of which lie deep in the crust and are difficult to gain access to.

The estimated overall chemical composition of the crust is shown in Table 1.1.

The continental plate material is sometimes referred to as sial and oceanic plate material as sima. Si + Al (=SIAL) are the chief components of granite. Si + Mg (=SIMA) are the main elements of the oceanic crust.

Table 1.1 Estimated percentage chemical composition of the Earth's crust

O	Si	Al	Fe	Ca	Na	K	Mg	Others
46.6	27.7	8.1	5.0	3.6	2.8	2.6	2.1	1.5

Below the crust is the **mantle**. This approximately 3000 km thick layer is believed to consist of dense rock (4.5 times the density of water) made up primarily of metal silicates. Its density however varies with depth and at least part of the outer mantle is fluid.

At the centre of the Earth is its **core**, with a radius of 3473 km. This is even more dense than the mantle (10.7 times the density of water) and is believed to be a mixture of iron and nickel, possibly with some sulfur and/or oxygen.

1.2 Terminology

Many disciplines have contributed to environmental science and each has brought with it specific terms. In describing the chemistry of the aquatic environment, it will be necessary to draw upon terms originating from fields such as biology, geology, hydrology, etc. Whilst it is not possible to cover all such terms, some of the most commonly encountered terms are described here. Additional explanations will be given throughout the text.

The spheres

The Earth is not just a geological mass, it brings with it an atmosphere, water, and life forms. For many studies it is convenient to consider these as distinct compartments. This book deals mainly with the **hydrosphere**, the collective name given to the many forms of water which are found on the Earth. This includes the oceans, lakes, streams, snowpack, glaciers, polar ice caps, and ground water. More than 70 per cent of the Earth's surface is covered by water and most of this is ocean.

The volume of ocean water is *c.* 1.36×10^9 km^3.

It is impossible to consider the hydrosphere in isolation as it is constantly interacting with the solid Earth, the atmosphere, and living organisms. These are covered by the lithosphere, atmosphere, and biosphere, respectively.

The lithosphere This is the outer part of the solid Earth. It is a cool rigid solid layer about 100 km thick made up of several rigid plates floating on the nearly molten part of the upper mantle (**asthenosphere**). The outer layer is **the crust**, the inner part of which is the edge of the upper mantle which has solidified due to the low temperature.

The atmosphere The envelope of gas surrounding the Earth. This is subdivided into different regions depending on altitude:

- Thermosphere 85–500 km above the Earth
- Mesosphere 50–85 km above the Earth
- Stratosphere 16–50 km above the Earth
- Troposphere 0–16 km above the Earth

The Biosphere This refers to life and includes living organisms and their immediate surroundings. The following descriptive terms are often used in association with biological processes.

Aerobic with air.

Ammonification the breakdown, by digestion and decay, of nitrogen-containing compounds to release urea and ammonia.

Anaerobic without air.

Anoxic without oxygen.

Autotrophic biota organisms which use sunlight or chemical energy to convert simple inorganic matter into the molecules of life.

Biota living organisms.

Decomposers heterotrophs which break down complex organic compounds to those which can eventually be used by autotrophic organisms (bacteria and fungi).

Epilimnion the warmer, less dense, surface water.

Eutrophication the excess growth of algae in very productive waters, which results in reduced dissolved oxygen levels due to the decomposition of dead algae.

Heterotrophic organisms live on organic compounds produced by autotrophic organisms.

Hypolimnion dense bottom water in, for example, a lake.

Nitrogen fixation conversion of dinitrogen to ammonia by aerobic and anaerobic micro-organisms.

Overturn the cooling of the epilimnion increasing its density causing a breakdown in water stratification.

Oxic with oxygen.

Photosynthesis the process by which plants convert carbon dioxide into sugars etc.

Phytoplankton photosynthesizing small drifting aquatic organisms (normally small aquatic plants).

Productivity the ability of a water body to produce living material.

Respiration the process by which an organism uses oxygen and organic matter as an energy source.

Salinity the ratio of the mass of dissolved salts in a solution to the mass of that solution.

Stratification layering.

Symbiotic mutually advantageous.

Thermocline the transition region between the epilimnion and hypolimnion in which temperature changes rapidly.

Zooplankton non-photosynthesizing small drifting aquatic organisms.

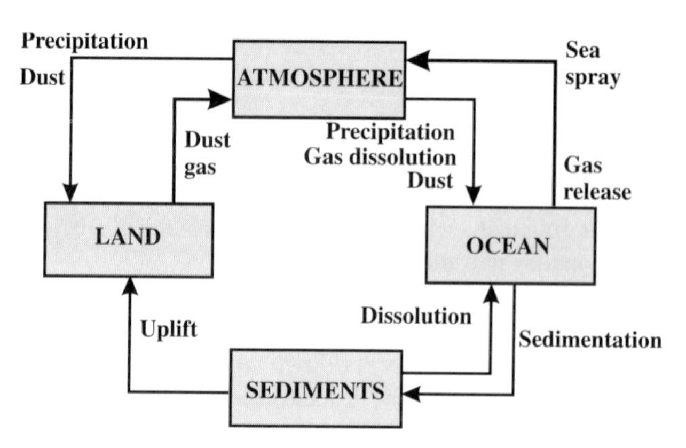

Fig. 1.2 A partial geochemical cycle.

1.3 Geochemical cycles

There can be little, if anything, on this Earth which remains unchanged with time. Material is constantly moving, changing both its physical and chemical form as a result of physical, biological, and chemical processes. Rocks are eroded by rivers and the resulting material is carried to the sea; plants take in carbon dioxide from the atmosphere, convert the carbon into biomass, and give out oxygen to the atmosphere. Human activities have led to perturbations of these natural fluxes of material but only rarely have they had any effect which could be considered to be significant on a global scale or in geological time.

This is not to say that global effects never happen, the thinning of the ozone layer and global warming are just two very important examples.

In order to show the movement of material through the environment, descriptive models are used. A geochemical cycle, for example, is a model which is used to describe the chemical aspects of the cycling of material through the geological environment (Fig. 1.2).

In such diagrams the boxes represent reservoirs of material joined by lines showing the movement (flux) of material between them. If the system is in a steady state, the average time that a component remains in a reservoir is called its residence time. This can simply be calculated by dividing the quantity of material in the reservoir by the rate at which it is added or removed:

$$\text{Mean residence time} = \frac{\text{amount of element in resevoir}}{\text{rate of addition/removal of element}}$$

The magnitudes of the residence times reflect how mobile the systems are; residence times in the crust are much longer than those in the sea, whilst atmospheric residence times are shorter. In seawater the residence times vary significantly depending on how the element is removed from the water (Table 1.2).

The amount of sodium in the ocean is *c.* 15×10^{18} kg and *c.* 220×10^{9} kg is added each year. This gives it a residence time of 68 million years.

The carbon cycle

Most of the Earth's carbon is tied up in rocks as inorganic carbonates (chalk and limestone, for example, are calcium carbonate) or as organic compounds in sedimentary rocks such as shales (Fig. 1.3).

The quantity of carbon in living organisms is only a small proportion of the Earth's carbon.

Less than 1 per cent of the Earth's carbon is to be found in the atmosphere, biota, soil, etc. Only a small proportion is therefore immediately available, and

Table 1.2 Mean residence times of the elements in seawater

Element	Residence time (10^6 years)	Loss mechanism
Na	68	Deposition of evaporites
Cl	100	Deposition of evaporites
Mg	10	Newly formed ocean crust
Ca	1	Shell formation
Pb	0.0004	Scavenging by particles
Al	0.0001	Adsorbed onto clays

Evaporite deposits: evaporation of isolated seawater results in the deposition of dissolved salts and the formation of rocks—limestone first, then more soluble constituents.

within our lifetime only a very small proportion will be processed into different chemical forms.

A similar situation exists in the sea. Although the oceans cover a large proportion of the Earth's surface, seawater absorbs and scatters light, stopping it penetrating more than about 200 m below the surface. There is little light below 200 m and this prevents photosynthesis.

The average ocean depth is 3000 m, and most of the carbon which is in the oceans is below the productive surface layer and is in the form of dissolved carbon dioxide and carbonate species.

The ocean is a sink for atmospheric carbon dioxide. CO_2 dissolves in the surface waters, is utilized by various organisms, the organisms subsequently die, falling through the water column to the bottom sediments.

An additional factor also limits the amount of living material in much of the World's oceans. Not only does light not penetrate to any great depth, but in many areas the nutrient elements which are required by photosynthetic organisms (nitrogen, phosphorus, silicon, etc.) are in short supply.

The most fertile areas of the oceans occur where rising (upwelling) currents bring nutrients to the surface. This occurs around a number of the continental coasts and the Antarctic seas.

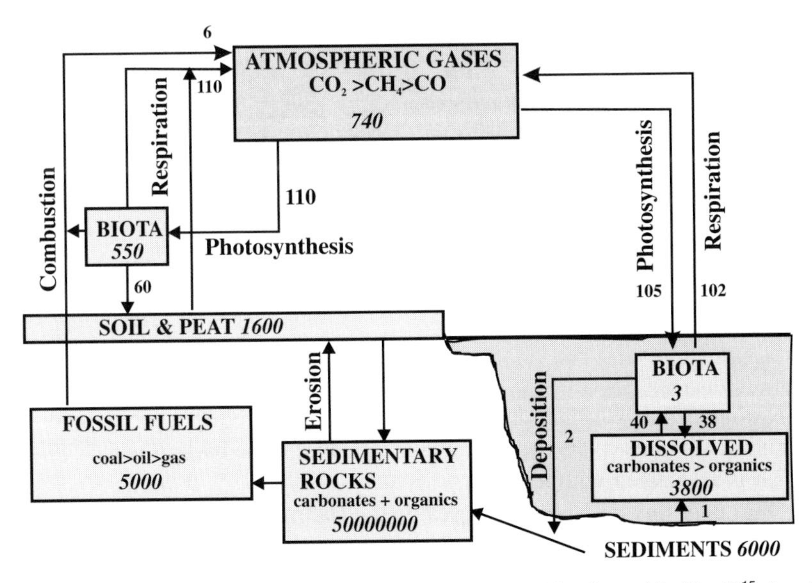

Fig. 1.3 The global carbon cycle showing the main reservoirs of the element (in Gt or 10^{15} g) and the estimated annual fluxes (in Gt per year).

When all these limitations come together the result is that whilst the oceans have a greater surface area than the land, the contribution of the oceans to the Earth's biomass is only about one half of one per cent.

The nitrogen cycle

Nitrogen is a widely distributed element found most commonly as nitrogen gas in the atmosphere, nitrate in soils and groundwaters, and, in biota, in the amino acids which make up proteins and the nucleotides of nucleic acids (Fig. 1.4).

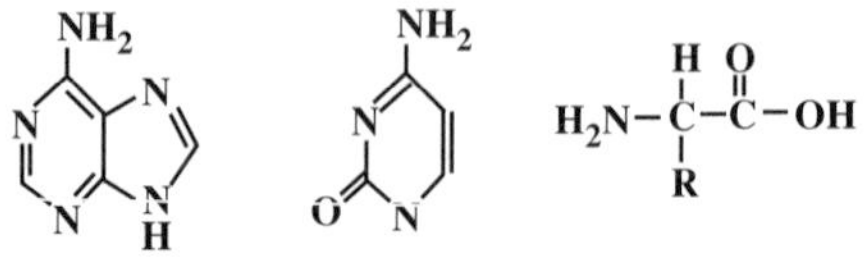

Fig. 1.4 Adenine, cytosine and amino acids (R is a variable structural feature).

The N–N bond strength in dinitrogen is 950 kJ mol^{-1} (very high).

Despite 78 per cent (by volume) of the atmosphere being gaseous dinitrogen (N_2), the availability of nitrogen from this source is restricted due to its low chemical reactivity.

Enzyme names are normally made by putting '-ase' at the end of a description of their function.

Atmospheric nitrogen cannot be directly utilized by most organisms and its availability is only made possible by its biological conversion to more chemically active forms, a process called **nitrogen fixation**. This is carried out by both aerobic and anaerobic micro-organisms (bacteria or blue–green algae) which convert dinitrogen to ammonia using the enzyme nitrogenase. The resulting ammonia can then be used by these and other organisms (Fig. 1.5).

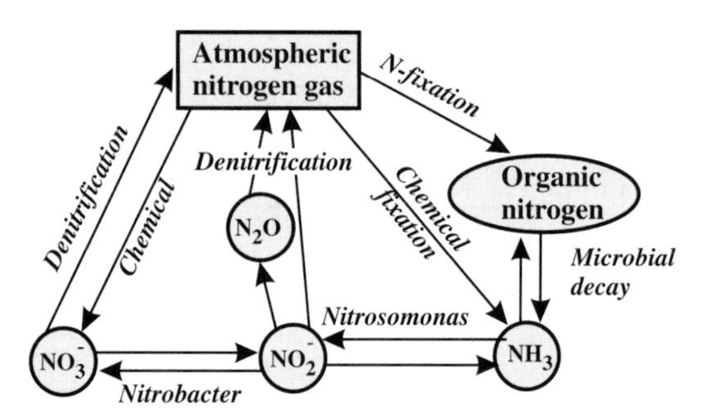

Fig. 1.5 The nitrogen cycle.

Legumes = peas, beans, etc.

Nitrogen fixing organisms occur both as free-living species and as symbiotic nitrogen fixers. An example of the latter is the bacterium *Rhizobium* which is associated with legumes.

Once over this initial barrier posed by the chemical inertness of dinitrogen, a range of biochemical processes convert the nitrogen into different forms

$$N_2 \rightarrow NH_3 \rightarrow NO_2^- \rightarrow NO_3^- \rightarrow \text{amino acids} \rightarrow \text{proteins}$$

The nitrogen-containing compounds found in plants and animals, when broken down by digestion and decay, are released back into the environment where they further break down to low molecular weight compounds such as urea and ammonia.

H_2N H_2N =O
Urea

How then does nitrogen return to the atmosphere as dinitrogen? This is performed by both aerobic and anaerobic micro-organisms in soil and the oceans. Some anaerobic organisms can use nitrate instead of dioxygen as an electron acceptor and energy source, leading to the generation of dinitrogen.

$$5(CH_2O) + 4NO_3^- + 4H^+ \rightarrow 2N_2 + 5CO_2 + 7H_2O$$

The conversion to dinitrogen is however incomplete and this leads to the formation of some nitrous oxide (N_2O), the second most abundant nitrogen-containing species in the atmosphere (0.3 ppm)

The mercury cycle

All elements in the environment take part in biogeochemical cycles. Whilst the carbon and nitrogen cycles are examples of the cycling of essential non-metals, such cycles also describe the movement of metallic elements, whether essential or polluting, through the environment. An example of an element which is currently believed to have no essential function is mercury (Fig. 1.6).

The element can be present in a number of chemical forms, the nature of which governs its behaviour in the environment. In solution its simplest form is the hydrated mercury(II) ion Hg^{2+}, but in the presence of large quantities of chloride ions (such as in the sea) it is complexed to give $HgCl_4^{2-}$. Changing from oxidizing to reducing conditions can result in the formation of metallic mercury (Hg^0) which, although soluble at low concentrations in water, has a significant vapour pressure which can result in the loss of the element to the atmosphere. In

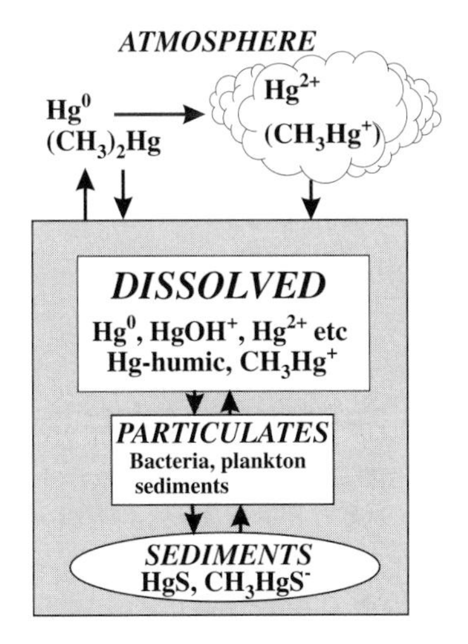

Fig. 1.6 The cycling of mercury through the environment.

the presence of the sulfide generated by bacteria in anoxic sediments, the mercury can be deposited into the sediment as the insoluble HgS species.

Biological organisms influence the environmental cycling of mercury in a number of ways. Firstly the element is stored in many organisms attached to a sulfur-rich protein called metallothionein. Much of the interest in this element has however arisen from the biological methylation of the element to give the highly toxic methylmercury (CH_3Hg^+), the chemical species which was responsible for a catastrophic poisoning incident in Minimata Bay (Japan) during the 1950s.

1.4 Problems

1.1 From where does a marine macro-alga (seaweed) obtain its carbon, nitrogen and sulfur?

1.2 Explain why the residence time of N_2 in the atmosphere is 1.6×10^7 years whilst the corresponding lifetime of ammonia is only 10 days

1.3 From the viewpoint of chemical changes induced in the enviroment, organisms are frequently regarded as 'black boxes' having inputs and outputs. For you and me, the main carbon inputs would be carbohydrates, fats, and proteins. The outputs would be carbon dioxide and waste products. Identify the carbon inputs and outputs for typical photosynthetic, methanogenic, and sulfate-reducing organisms.

1.5 Further reading

Global biogeochemical cycles. Butcher, S. S., Orians, G. H., Charlson, R. J., and Wolfe, G. V. Academic Press, 1982. ISBN 0-12-147686-3

Earth. Press, F. and Siever, R. 4th Edition, Freeman, 1985. ISBN 0-7167-1743-3

Biogeochemistry; An analysis of global change. Schlesinger, W. H. Academic Press, 1997. ISBN 0-12-625155-X

Earth science. Tarbuck, E. J. and Lutgens, F. K. Prentice Hall, New Jersey, 1997. ISBN 0-13-570839-7

Modern physical geology. Thompson, G. and Turk, J. Saunders College Publishing, 1996. ISBN 003-005222-X

Limnology. Wetzel, R. G. Saunders College Publishing, 1982. ISBN 003-057913-9

1.6 General texts on environmental chemistry

Chemical principle of environmental pollution. Alloway, B. J. and Ayres, D. C. Blackie, London, 1997. ISBN 0-7514-0380-6.

An introduction to environmental chemistry. Andrews, J. E., Brimblecombe, P., Jickells, T. D. and Liss, P. S. Blackwell Science, Oxford, 1996. ISBN 0-632-03854-3.

Environmental chemistry. Baird, C., W. H. Freeman & Co, New York, 1995. ISBN 0-7167-2404-9.

Chemistry in your environment, user-friendly, simplified science. Barrett, J. Albion Publishing, Chichester, 1994. ISBN 1-898563-01-2.

An introduction to environmental chemistry. Bunce, N. J. Wuertz Publishing Ltd, Winnipeg, 1993. ISBN 0-920063-50-0.

Understanding our environment: an introduction to environmental chemistry and pollution. Harrison, R. M. Royal Society of Chemistry, Cambridge, 1992. ISBN 0-85186-233-0.

Global environmental issues, a climatological approach. Kemp, D. D. Routledge, London, 1992. ISBN 0-415-01108-6.

Environmental chemistry. Manahan, S. E. Lewis Publishers, Chelsea Michigan, 1995. ISBN 1-56670-088-4.

Environmental chemistry. O'Neill, P. Chapman and Hall, London, 1993. ISBN 0-412-48490-0.

2 The aquatic environment

2.1 Types of natural waters

There are many ways by which natural waters can be classified and it must be recognized that the movement of water through the environment does not limit it to any single category. It is often convenient however to consider the Earth's surface as a suitable cut-off point.

Groundwater

This is the water below the Earth's surface. The interface between the upper region, where air is present in the soil and underlying rocks, and the level below which any pores, cracks etc. are filled with water is called the **water table**. The quantity of water that can be held as groundwater and its mobility are dependent upon the nature of the geology. The quantity of water that can be held within the rock structure is governed by its porosity.

porosity = the proportion of a rock which is void space and which can hold water.

A rock which is porous does not necessarily permit the ready movement of water through it and the term **permeability** is used to describe the ability of water to move through the rock. The size, nature, and interconnections between the pores, voids, and cracks through a rock governs its permeability.

Limestones often have low porosities but high permeabilities. This results from the rock having large channels and fissures. A shale on the other hand may hold a lot of water due to its high porosity, but permeability through its small pores may be slow.

Groundwater spends a significant time in contact with the underlying rock and this results in the dissolution of minerals and nutrients. The chemical composition of the groundwater will therefore reflect the underlying geology of the region. As it moves through the rock beds much of the dissolved and particulate material which has been picked up at the surface is removed by filtration and adsorption but at the same time some of the rock material will dissolve. Where the water table intercepts the surface, groundwater seeps out forming springs, streams, swamps, and lakes.

Surface waters

The concentration of salt in the water allows us to divide surface waters into two broad categories. Freshwater is distinguished from saline water by its low salt content and is most commonly found in rivers and lakes.

Local situations, such as volcanic activity, can result in extreme natural water conditions not covered in this general description.

The most significant example of a saline water is oceanic seawater. Whilst the chemical composition of seawater varies slightly from area to area due to the addition of freshwater and evaporation, it typically contains *c.* 35 g of dissolved salt per litre. This is made up of a number of different ions, the most dominant of which are sodium and chloride.

Freshwater and seawater come together in estuaries resulting in regions of intermediate salt content. In a well-mixed estuary a salinity gradient occurs between the river and the estuary mouth (Fig. 2.1).

Freshwater and seawater have different densities and in many poorly mixed estuaries the less dense river water can lie on top of the incoming seawater for a large proportion of the estuary; such an estuary is said to be stratified.

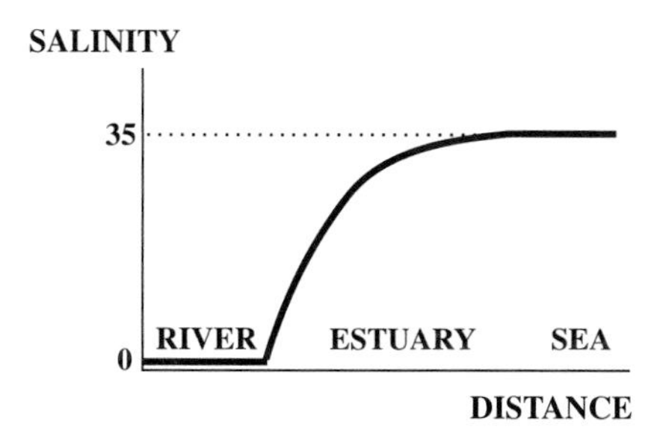

Fig. 2.1 The salinity (S) along a typical estuary.

Salinity is the weight in grams of inorganic ions dissolved in 1 kg of the water. In practice this cannot be measured and it is normally assessed by measuring the conductivity of the water relative to the conductivity of a standard seawater. This measure of salinity is therefore unitless. Older texts however will be found which give measured salinity values with units of parts per thousand (1‰) or g dm^{-3}. Typical open ocean salinities lie in the range 32 to 37.

Estuaries are complex areas in which both dissolved and particulate materials are subjected to often quite rapid changes in chemical and physical environment. Most significant of the changes that occur are those involving pH and salinity. If either of these change significantly, it can induce the precipitation of dissolved species or the redissolution of material from the sediments. Those elements which are neither added to, nor removed from the dissolved phase during the mixing of freshwater and seawater in the estuary are described as behaving **conservatively** (Fig. 2.2).

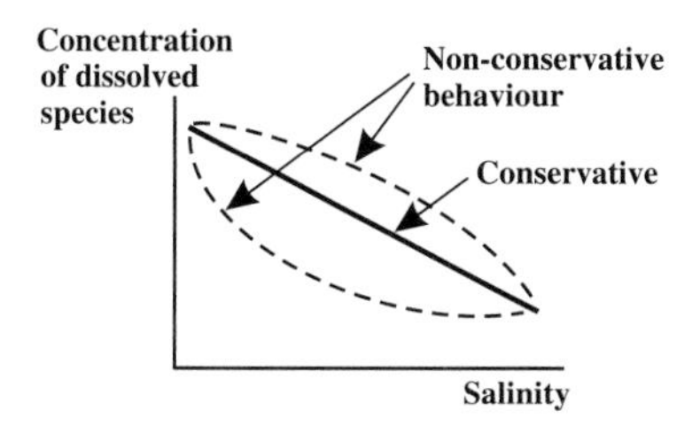

Fig. 2.2 Conservative and non-conservative behaviour of dissolved material in an estuary.

This is normally reflected in the concentration of a dissolved component being linearly related to salinity. **Non-conservative** behaviour results from material being deposited or added to solution during its movement through the estuary.

A significant example of such behaviour is the precipitation of riverine colloidal iron as it encounters the increasing salinity in an estuary. This results in iron being deposited into the sediments.

The precipitating material may take with it other elements by co-precipitation or adsorption.

2.2 The hydrologic cycle

Whilst much of the World's water is to be found in the oceans, it is also present in lakes, rivers, snowpacks, glaciers, polar ice caps, in the ground, organisms, and the atmosphere. It is constantly moving, both as liquid and vapour, transporting with it the gases and nutrient materials required for life. Its movement is a cycle, recirculating water between temporary storage compartments (Fig. 2.3).

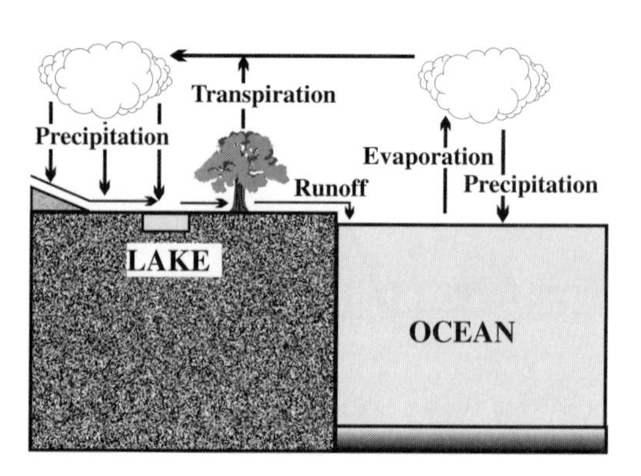

Fig. 2.3 The cycling of water.

Transpiration = loss of water from plants.

This cycling is not limited to purely physical processes. Vegetation cover, for example, can reduce the rate of direct evaporation of water from the soil, but at the same time will result in the loss of water to the atmosphere by **transpiration**.

2.3 Water, nature's solvent

Light only penetrates to about 200 m in the oceans, limiting photosynthesis to a relatively small upper layer of water.

Water is the most ubiquitous liquid on the Earth's crust and it has many properties which govern its environmental impact. Firstly, it has a very high dielectric constant making it an excellent solvent for ionic substances. It therefore causes the dissolution of minerals, increasing their mobility and aiding the delivery of nutrients to plants. For aquatic plants to survive, light must be available to them for photosynthesis. Water is transparent to visible and long wavelength ultraviolet light permitting the penetration of the light required for photosynthesis to some depth.

The ability of water to store heat, and the energy required for it to evaporate, play important roles in maintaining the structure and temperature of water bodies. It is relatively difficult to alter the temperature of a mass of water due to the high heat capacity of water (4.19 kJ kg^{-1} K^{-1}). Water also has an extremely high heat of vaporization (2260 kJ kg^{-1}) which limits its vaporization, protecting the organisms which live in it.

The density–temperature relationship of water is rather unusual (Fig. 2.4) and has a number of environmental consequences. The density of water is at a maximum at 4 °C above its freezing point.

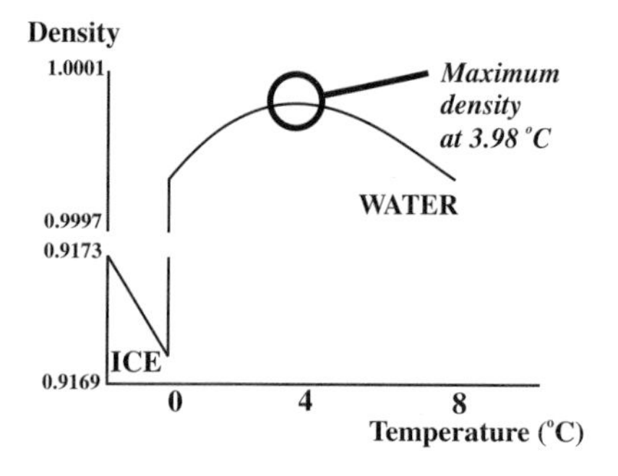

Fig. 2.4 Variation of water density with temperature.

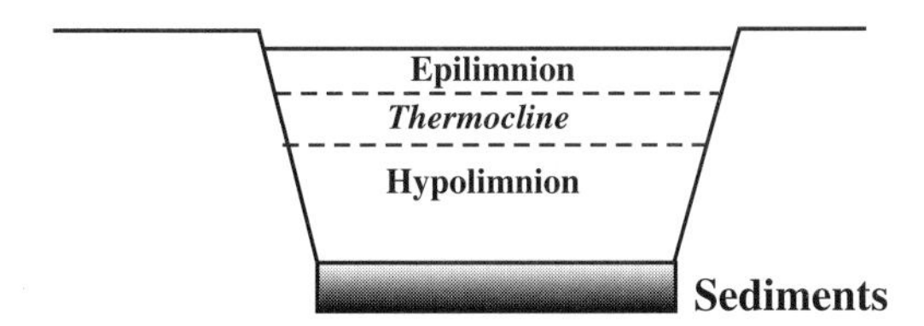

Fig. 2.5 The stratification of a lake.

This causes ice to float and prevents large bodies of water from completely freezing. In non-flowing bodies of water, such as lakes, stratification can occur due to the solar heating of the water surface. The warmer water (the **epilimnion**), having a lower density, floats on a cooler (more dense) bottom layer (the **hypolimnion**) (Fig. 2.5).

Between these two layers there is a region in which the temperature changes rapidly called the **thermocline**. One consequence of this thermal stratification and poor mixing between the layers is that whilst the epilimnion is in contact with the atmosphere, and is therefore well provided with oxygen, the hypolimnion is cut off from an oxygen source and can become depleted of oxygen. This depletion can become particularly pronounced as the little oxygen which is present is used up in oxidizing dead organic matter falling to the bottom of the lake. The upper levels are therefore favourable to **aerobic** organisms, often resulting in a heavy growth of algae.

Aerobic = requiring free oxygen for respiration.
Anaerobic = not requiring oxygen.

If the lower water becomes oxygen depleted this will favour the growth of **anaerobic** organisms such as **methanogenic** and **sulfate-reducing** bacteria.

Methanogenic = methane producing. Sulfate-reducing bacteria obtain their energy by reducing sulfate (S(VI)) down to sulfide (S(−II)).

During the period of stratification nutrients build up in the bottom waters as they are released from decaying matter. In the autumn the water stratification becomes unstable as the epilimnion cools causing the surface water to increase in density and to mix with the lower water. As this happens the thermal stratification disappears (the **overturn**). Chemicals in the water layers mix and the water body becomes more uniform, providing a release of nutrients from the lower water into the nutrient-deficient upper water which induces an increase in biological activity. The fertilization of the upper water column results in a burst of growth of micro-algae called a phytoplankton bloom. These blooms commonly occur in both the spring and autumn as a result of the overturns.

Phytoplankton are free drifting photosynthetic aquatic organisms ranging in size from less than 1 μm to greater than 0.2 mm.

2.4 Dissolved inorganic compounds in natural waters

The quantity of dissolved material in natural waters varies greatly. At the two common extremes, rainwater may contain as little as a few milligrams of dissolved material per litre whereas seawater contains approximately 35 g of salt per litre. The presence of even a few milligrams of dissolved matter in a litre of rainwater may at first seem surprising. This is derived from a number of sources: sulfate from the oxidation of hydrogen sulfide, sulfur dioxide, and dimethylsulfide, and nitrogenous material from terrestrial vegetation and

Dimethylsulfide (Me_2S) comes from the breakdown of the dimethylsulfoniopropionate generated by some phytoplankton in the World's oceans. Being a volatile species it is released into the atmosphere.

combustion emissions. Especially in coastal regions, a large proportion of the material is derived from sea salt in aerosol particles released into the atmosphere from breaking waves.

After falling onto the land, rainwater moves both through and over the soil and underlying rocks dissolving material as it travels. Organic matter is picked up during this process but can be removed from the water if it subsequently travels through the groundwater system. Such a route may also increase the inorganic content of the water.

The concentrations of metals which are present in natural waters are highly variable, reflecting the geology of the catchment area and its history. A large proportion of the material which is dissolved in water is in fact made up of metal salts (Table 2.1).

You are probably aware that the concentration of hydrogen ions in solution can be expressed as pH but may not have experienced other concentrations expressed as p-functions. In general terms:

$$pX = -\log_{10}[X]$$

Remember that the higher the value of pX, the lower the concentration of X. [1 pH unit corresponds to a 10-fold change in the hydrogen ion concentration.]

Table 2.1 The compositions of some typical 'freshwater' bodies

	Streams	Groundwater	Lake	Soda lake
pH	6.6–8.0	7.0–8.0	7.7	9.6
pNa	4.0–4.6	2.6–3.5	3.4	0
pK	4.7–5.1	3.7–4.5	4.3	1.7
pCa	3.1–4.3	2.5–3.5	3.0	4.5
pMg	4.0–5.1	2.5–3.8	3.4	4.6
pH_4SiO_4	3.8–4.2	3.0–3.9	4.7	2.8
$pHCO_3$	2.9–4.0	2.1–2.9	2.7	0.4
pCl	5.3–5.8	3.2–4.0	3.6	0.3
pSO_4	3.7–4.7	2.2–4.7	3.6	2.0

At the other extreme, seawater contains *c.* 3.5 per cent by weight of dissolved salts, the most significant of which is sodium chloride. Just six ions account for over 99 per cent of the salinity of seawater (Table 2.2).

Table 2.2 The major constituents of seawater

Ion	Percent by weight
Cl^-	55.07
Na^+	30.62
SO_4^{2-}	7.72
Mg^{2+}	3.68
Ca^{2+}	1.17
K^+	1.10
Total	99.36

2.5 Organic material in natural waters

All natural waters contain organic matter. Some of this is living, ranging in size from bacteria to whales; some is decaying, and some has been introduced from external sources such as drainage, factory pollution, and atmospheric fallout. The natural dissolved organic matter in water is made up of the transformation products of **biogenic** material and biological excretion products. Organic compounds are secreted and excreted by all living organisms and these in turn are broken down into smaller species by bacteria and chemical processes. Proteins, amino acids, fats, lignins, etc., the structural elements of plants and animals, are being broken down continuously both whilst the organism is living and as it decays. The products of the degradation and the compounds which are formed by the reassociation of these fragments, lead to a complex mix of dissolved organic compounds in the water.

The dissolved organic matter (DOM) in seawater ranges from 0.1 to 1 mg/litre and up to *c.* 10 mg/litre in lakes and rivers.

This organic matter covers the full range of molecular weight, ranging from totally soluble low molecular weight organics to high molecular weight polymers of colloidal and particulate nature.

Small organic compounds

In the absence of pollution sources, small organic molecules such as amino acids, sugars, fats, and chlorophyll are derived from animal and plant metabolism and the decomposition of larger biologically derived molecules (Fig. 2.6). These in turn are a major source of nutrition for other organisms. In

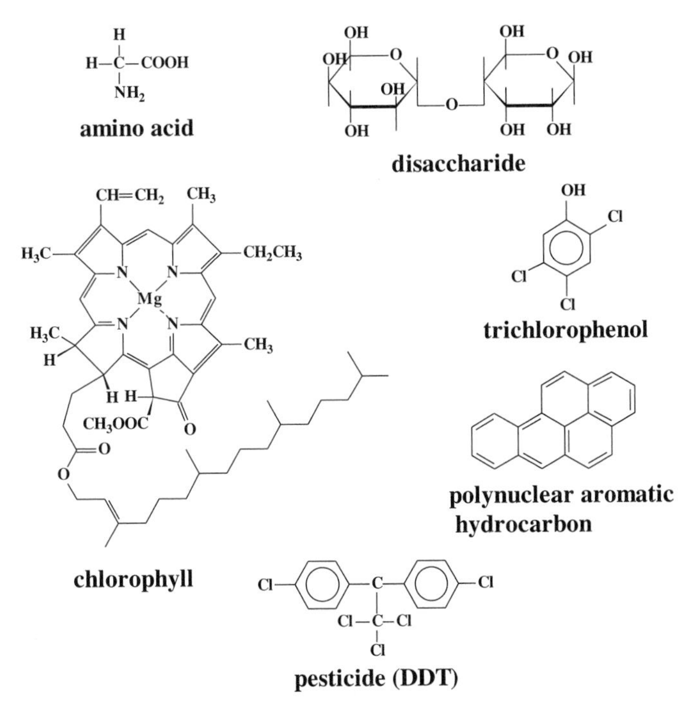

Fig. 2.6 Some of the smaller organic molecules found in natural waters.

addition to the naturally occurring compounds there are many which are present as a result of man's activities, examples of which include pesticides such as DDT and the polynuclear aromatic hydrocarbons resulting from combustion.

Thousands of compounds have so far been identified, some of which, due to their size and charge, remain in the water. Some adsorb onto the surface of particles and other more volatile species partition into the atmosphere.

Higher molecular weight compounds

Humic matter gives the brown coloration associated with peaty water.

Much of the dissolved organic matter is of high molecular weight and falls under the title of humic matter.

Humic matter is a complex mixture of polymeric material, having molecular weights above 300 Dalton. Its solubility and metal complexing abilities are largely determined by it containing a large number of phenolic and carboxylic acid groups. Historically, this material has been divided into three main groups according to the conditions under which the components can be extracted. These are:

- **humic acids**—compounds which are alkali soluble but which are precipitated by acid
- **fulvic acids**—lower molecular weight compounds which are soluble at all pH
- **humin**—material which cannot be extracted by acid or base.

Humic material has no single structure and it is only therefore possible to make generalizations regarding its structure from NMR, infrared, group specific colorimetry, and titrimetric analysis. Due to its ready availability and the ease by which significant quantities of humic matter can be extracted, more is known about the humic matter that can be extracted from peat and river waters than any other source. Marine humic matter is currently believed to be much less aromatic in structure than its freshwater equivalent. Figure 2.7 illustrates some of the structural features which have been proposed as typifying humic matter.

NB This is only a representative structure as no two 'molecules' of humic matter are likely to be identical and an almost infinite number of structures is possible.

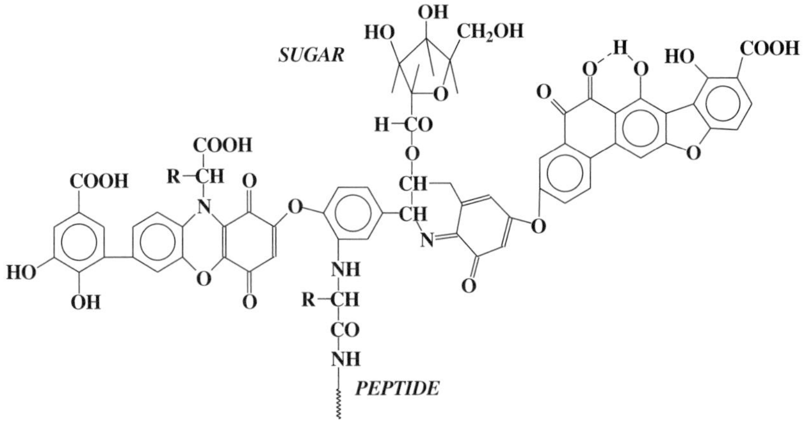

Fig. 2.7 An example structure showing some of the features which are characteristic of dissolved humic substances found in freshwaters.

2.6 Summary

Water is the most important liquid on the Earth and has a major impact on the chemical, physical, and biological processes which take place. Table 2.3 summarizes some of the unusual characteristics of water and the resulting environmental consequences.

Table 2.3 Properties of water

Property	Magnitude	Consequence
Heat capacity	Exceptionally high ($4.19\ kJ\ kg^{-1}\ K^{-1}$)	a) Slows down temperature changes. b) Heat transported around the globe by ocean currents. c) Influences climate.
Latent heat of fusion	Exceedingly high ($333\ kJ\ kg^{-1}$)	Stops the water temperature from changing rapidly when it is around 0 °C due to the additional energy required to freeze or thaw the water.
Latent heat of evaporation	Highest of all substances ($2260\ kJ\ kg^{-1}$)	Cuts down water and heat loss to the atmosphere.
Density	ca. $10^3\ kg\ m^{-3}$. Maximum density at 4 °C. Decreases with increasing salinity	Ice floats, insulating the water below from cold atmosphere.
Surface tension	Highest of all liquids ($73\ mN\ m^{-1}$)	Controls the shape of raindrops, sea spray, etc.
Dissolving power	Exceptionally good	Dissolves nutrients and transports them to plants.
Transparency	Relatively large	Absorbs in the ultraviolet and infrared parts of the spectrum but transmits the visible radiation required for photosynthesis.

2.7 Problems

2.1 Use the information provided in Section 2.4 to calculate the pCl of a typical sample of seawater.

2.2 Seawater contains approximately 3.5% by weight dissolved solids. Use the data in Table 2.1 to estimate the concentration of dissolved solids in typical lake water.

2.3 Much of the humic matter to be found in streams and rivers is derived from lignin. What is lignin and where does it come from?

2.4 Comment on the observation that rainwater collected from a coastal region was found to contain, on a weight basis, almost twice as much dissolved chlorine as sodium.

2.8 Further reading

Principles of hydrology Ward, R. C. and Robinson, M. 3rd Edition, McGraw Hill, Maidenhead, 1989. ISBN 0-07-707204-9

An introduction to organic geochemistry. Killops, S. D. and Killops, V. J., Longman Scientific, Harlow, 1993. ISBN 0-582-08040-1

Seawater: its composition, properties and behaviour. The Open University Course Team, Pergamon, 1985. ISBN 0-08-0425186

Land, water and development. Newson, M., Routledge, London, 1995. ISBN 0-415-05711-6

An introduction to marine biogeochemistry. Libes, S. M. Wiley, New York, 1992. ISBN 0-471-50946-9

3 The acidity of water

3.1 Introduction

In recent years there has been much publicity over acid rain, a problem which normally results from the dissolution of atmospheric pollutants such as sulfur dioxide and the oxides of nitrogen.

Acid precipitation can occur as both wet and dry deposition—it is not in fact limited to rain.

In some areas the acid nature of the local precipitation has major effects on vegetation and fish stocks, in other areas there appears to be little impact. Why is there such a difference?

The acidity of a natural water depends on the nature of material dissolved in it and its interaction with other materials such as the rocks, the organisms living in it, and added pollutants. Rivers and lakes in granite areas, for example, are unable to neutralize any added acidity and are therefore highly susceptible to the effects of acid rain. Significant increases in the acidity of these water bodies produces an environment which is not well suited to life for a number of reasons. Firstly, most organisms are not well adapted to acidic conditions or changes in acidity. In addition, acidification leads to the dissolution of a number of toxic elements which can poison flora and fauna. In limestone areas, however, the water can neutralize moderate quantities of added acid. Why this comes about and how this affects other species in solution is one of the subjects covered by this and subsequent chapters.

Granite is a hard, coarse-grained igneous rock which is highly resistant to weathering. It is composed of grains of several minerals including feldspar and quartz.

3.2 The chemical nature of water

Water is a mixture of three main components: H_2O and its dissociation products the hydronium ion H_3O^+ and the hydroxide ion OH^-.

$$2H_2O \rightleftharpoons H_3O^+ + OH^-$$

The bare proton H^+ is very small and has a very high charge density. It is very strongly attracted to the negative end of the water dipole. In aqueous solution the proton is hydrated and is more accurately represented as H_3O^+.

When nothing else is present in the water the concentrations of H_3O^+ and OH^- are equal. If anything is added to the water which has an affinity for H_3O^+ ions (such as ammonia), the concentration of H_3O^+ ions drops and the solution becomes less acidic (i.e. becomes alkaline).

3.3 The acidity of water

The acidity of any aqueous solution is a measure of the concentration of hydronium ions in solution; it is measured in terms of pH

$$\text{pH} = -\log_{10}[H_3O^+]$$

This simple little equation raises a number of points which provide the basis of more advanced treatments of thermodynamic equilibria. In order to simplify the situation we will have to make a few assumptions:

1. The effective concentration of a species in solution is not always the same as the concentration which is measured out due to interactions between the species in solution. This leads to the concept of activity in which activity (a) is the measured concentration (c) multiplied by an activity coefficient (γ).

$$a = \gamma c$$

For dilute solutions the activity and concentration can be considered to be the same ($\gamma = 1$). This is not however the case for seawater (ionic strength *c.* 0.6 mol dm^{-3}) in which activity coefficients can fall to below 0.1. This introductory text will assume that concentrations can be used to represent the activities of species in solution.

2. You may have noticed that we come across a problem here when it becomes necessary to take the logarithm of a unit. This is not mathematically possible and more formally the equation should be written:

$$\text{pH} = -\log_{10}\left(\frac{[\text{H}_3\text{O}^+]}{[\text{H}_3\text{O}^+]^{\ominus}}\right)$$

where $[\text{H}_3\text{O}^+]^{\ominus}$ is the standard state and is normally 1 mol dm^{-3}

Throughout this book concentrations will be expressed relative to this implied standard state, thereby overcoming the problems involved with the associated units.

The definition of pH then becomes:

$$\text{pH} = -\log_{10}[\text{H}^+]$$

3. Whilst recognizing that the species in solution might be better represented as H_3O^+ it is very common practice to write H^+ instead of H_3O^+, as this simplifies equations.

In subsequent equations this convention will be employed. Written in its simpler form the dissociation of water is described by the equilibrium:

$$\text{H}_2\text{O} \rightleftharpoons \text{H}^+ + \text{OH}^-$$

The equation represents two reactions at the same time.

The forward reaction is the breakup of H_2O into H^+ and OH^-. The Gibbs energy change associated with this reaction is ΔG_F.

$$\text{H}_2\text{O} \rightarrow \text{H}^+ + \text{OH}^-$$

At the same time a reverse reaction is occurring in which OH^- and H^+ combine to give water, resulting in a free energy change of ΔG_R.

$$\text{H}^+ + \text{OH}^- \rightarrow \text{H}_2\text{O}$$

At equilibrium the Gibbs energy of the system must be at a minimum otherwise there would be a driving force for the reaction to proceed in the forward or reverse direction.

As this is an equilibrium, the concentration of any species in solution can not be changing with time. The forward and backward reactions must be in balance with an overall Gibbs energy change for the combined reactions of zero.

The extent to which the water is dissociated is given by the equilibrium constant K:

$$K = \frac{[H^+][OH^-]}{[H_2O]} = 1.8 \times 10^{-16} \text{ at } 25\ °C$$

The square brackets denote the concentration of the species within the brackets relative to the standard state (1 mol dm^{-3}) and the equilibrium constants are therefore unitless.

The equilibrium constant for the reaction is very small indicating that only a small proportion of the water is ionized. The concentration of undissociated water ($[H_2O]$) will therefore be essentially unaffected by any dissociation. As $[H_2O]$ is constant at *c.* 56 mol dm^{-3}, the equilibrium constant can therefore be simplified:

The concentration of water = the mass of 1 dm^3 of water divided by the relative molar mass of water, = 1000/18 = 55.56 mol dm^{-3}

$$K = \frac{[H^+][OH^-]}{55.56}$$

$$K_W = [H^+][OH^-] = K \times 55.56$$

$$= 55.56 \times 1.8 \times 10^{-16}$$

$$K_W = 1 \times 10^{-14} \text{ at } 25\ °C$$

This is temperature dependent.

The new constant, K_W, is called the ionic product of water. In pure water equal concentrations of H^+ and OH^- are formed from the dissociation of the water and their concentrations must therefore be equal—the solution is described as being neutral.

Neutrality occurs when $[OH^-] = [H^+]$, i.e. when $K_W = [H^+]^2$ or $[H^+] = 10^{-7}$ mol dm^{-3} (pH 7).

The ionic product of water can be re-written in terms of pH and pOH:

$$pK_W = pH + pOH$$

$K_W = [H^+][OH^-]$
$\log K_W = \log[H^+] + \log[OH^-]$
$-\log K_W = -\log[H^+] - \log[OH^-]$
$pK_W = pH + pOH$

pK_W has been defined in the same way as pH and is equal to $-\log_{10}(K_W)$. As $K_W = 10^{-14}$, $pK_W = 14$ and pOH = 14 − pH.

Calculating the acidity of a solution

The calculation of the acidity of a solution is a suitable example with which to introduce the methods which are employed in all equilibrium calculations. The objective of most equilibrium calculations is to calculate the concentration of one particular species from known concentrations, constants, etc. This can sometimes be a daunting task, but if a stepwise approach is taken the process can be significantly simplified.

Step 1. *Know what species might be in solution.*

A solution of sodium chloride in water contains five species: H_2O, H^+, OH^-, Na^+, and Cl^-; the sodium chloride has been completely dissociated.

Step 2. *Identify how the concentrations of the species depend on each other.* Let us now take the example of an aqueous solution of a weak acid, such as acetic (ethanoic) acid.

The chloride ion has very little tendency to pick up a proton to give hydrochloric acid (hydrochloric acid is a strong acid) or to ion-pair with the sodium. Similarly the sodium ion has no affinity for the only other negative ion, OH^- as this would give rise to the strong base NaOH (which completely dissociates in solution!).

Acetic acid (HAc) is a weak acid. Its dissociation in solution is therefore incomplete and can be described by the acid dissociation constant:

$$CH_3COOH \rightleftharpoons CH_3COO^- + H^+ \quad K_a = 10^{-4.75}$$

$$\text{HAc} \rightleftharpoons \text{Ac}^- + \text{H}^+$$

$$K_a = \frac{[\text{Ac}^-][\text{H}^+]}{[\text{HAc}]}$$

$K_W = [H^+][OH^-]$

There is an additional equilibrium occurring in the aqueous solution which must never be forgotten. This is that $[H^+]$ and $[OH^-]$ are related by K_W.

Step 3. *How do these concentrations relate to the total amount of material put into solution?*

This is really a stock-taking exercise. When 1 mole of acetic acid is placed in a bucket full of water, it has nowhere else to go. Irrespective of whether any one molecule of acetic acid dissociates or not, the 1 mole of acetate must remain in the bucket. If the mass of acetic acid dissolved in the water were m_{Ac}, the *mass balance equation* could be written:

In words: 'the mass of acetate added (m_{Ac}) equals the sum of the masses of acetate which are present as undissociated acetic acid and as the acetate ion'.

$$m_{\text{Ac}} = m_{HAc} + m_{Ac^-}$$

As the volume of water (V) is fixed, we can convert the mass balance equation to its concentration form:

Throughout we will be using C to represent the 'analytical' concentration, i.e. the concentration of the material which would be physically made up in the laboratory.

$$m_{\text{Ac}} = m_{\text{HAc}} + m_{\text{Ac}^-}$$

$$\frac{m_{\text{Ac}}}{V} = \frac{m_{\text{HAc}}}{V} + \frac{m_{\text{Ac}^-}}{V}$$

$$C_{\text{Ac}} = [\text{HAc}] + [\text{Ac}^-]$$

$$C_{\text{Ac}} = [\text{CH}_3\text{COOH}] + [\text{CH}_3\text{COO}^-]$$

Step 4. *The charge balance*. Just as we have had to account for all the material added to our bucket, we must also ensure that the charges on the positive species equal the number of negative charges. This leads us to the *charge balance equation*.

If the positive and negative charges in our bucket full of acetic acid were out of balance, we would get a shock when handling it.

For sulfuric acid in solution:

In words: 'the sum of the positive charges equals the sum of all the negative charges'.

$$[\text{H}^+] = [\text{OH}^-] + [\text{HSO}_4^-] + 2[\text{SO}_4^{2-}]$$
$$+\text{ve charges} = -\text{ve charges}$$

Try not to forget the OH^- and H^+ from the water and to account for whether there are 1, 2, or 3 charges on each ion.

Note how each SO_4^{2-} contributes two charges and a factor of 2 is therefore placed in front of its concentration, H^+, OH^-, and HSO_4^- only contribute one charge each.

Let us now follow these steps to provide the information necessary to calculate the pH of a simple solution of hydrochloric acid (10^{-3} mol dm^{-3})

Step 1. *Identify the species in solution*. There are just four species in solution:

$$\text{H}_2\text{O},\ \text{H}^+,\ \text{OH}^-,\ \text{and Cl}^-$$

Step 2. *How do the concentrations of these species depend on each other?* The hydrochloric acid is a strong acid and it therefore completely dissociates in solution. We must not however forget the dissociation of the water, for which:

$$H_2O \rightleftharpoons H^+ + OH^-$$
$$K_W = [H^+][OH^-] = 10^{-14} \qquad (3.1)$$

Step 3. *Mass balance for chlorine*: If we knew nothing about the extent to which HCl dissociates in water we would have to assume that the chlorine ended up as both Cl^- and HCl and we would write:

$$C_{HCl} = [Cl^-] + [HCl]$$

to account for both possible chlorine species. There is nothing fundamentally wrong with this equation, it is just that we know that hydrochloric acid is a strong acid and therefore under non-extreme conditions [HCl] will be small.

Even for a 'strong acid' if the solution were extremely acidic the concentration of HCl could become significant.

The mass balance equation therefore simplifies to:

$$C_{HCl} = [Cl^-] \qquad (3.2)$$

Step 4. *Charge balance*. The only charged species in solution are H^+, OH^-, and Cl^-. The total positive charge must be the same as the total negative charge, giving:

$$[H^+] = [OH^-] + [Cl^-] \qquad (3.3)$$

Step 5. We now have a number of equations relating the concentrations of H^+, OH^-, Cl^-, and H_2O to each other. These are simultaneous equations which can now be solved to give us the concentration of any one species. In this particular example we want to know $[H^+]$ as this is pH. We must therefore eliminate everything except the required variable ($[H^+]$) and the terms for which numerical values are known (i.e. C_{HCl}).

Starting from eqn 3.3, eqn 3.1 can be used to convert $[OH^-]$ into $[H^+]$ and eqn 3.2 converts $[Cl^-]$ into C_{HCl}, a known concentration. These two substitutions give:

$$[H^+] = \frac{K_W}{[H^+]} + C_{HCl}$$
$$[H^+]^2 - C_{HCl}[H^+] - K_W = 0$$

This is a quadratic equation which can be readily solved to give $[H^+]$ and hence pH.

For the equation

$$ax^2 + bx + c = 0$$

the solution is

$$x = \frac{-b \pm \sqrt{b^2 - 4ac}}{2a}$$

$$[H^+] = \frac{C_{HCl} \pm \sqrt{C_{HCl}^2 + 4K_W}}{2}$$

$$= \frac{10^{-3} \pm \sqrt{10^{-6} + 4 \times 10^{-14}}}{2}$$

$$= 10^{-3} \text{ mol dm}^{-3}$$

$$pH = -\log_{10}[H^+] = 3$$

As hydrochloric acid is a strong acid, 10^{-3} mol dm^{-3} acid gives a proton concentration of 10^{-3} mol dm^{-3}, or a pH of 3.0

This has been a rather extended route to calculating the pH of 10^{-3} mol dm^{-3} hydrochloric acid. The simpler approaches, however, break down in more complex examples and the stepwise approach described here can, if necessary, be employed in quite complex calculations.

3.4 Polyprotic acids

To identify the number of acidic protons in many simple inorganic acids it is useful to rewrite their chemical formulae in an unconventional but informative form, grouping together the oxygen atoms according to whether they are terminal oxygens or part of an OH group. The following are examples:

$HNO_3 = N(V)(O)_2(OH)$
$H_2SO_4 = S(VI)(O)_2(OH)_2$
$H_3PO_4 = P(V)(O)(OH)_3$

Thus nitric acid is monoprotic, sulfuric acid diprotic, and phosphoric acid triprotic.

$H_2CO_3 = C(IV)(O)(OH)_2$ therefore diprotic.

Whilst acetic, hydrochloric and nitric acids have only one acidic proton in their structure which can be given up in aqueous solution, and hence have only one acid dissociation constant, many acids have more. These acids are called polyprotic acids.

One of the most important acids present in the aquatic environment is carbonic acid. This arises from the dissolution of carbonate rocks such as chalk and limestone, and from the dissolution of carbon dioxide generated by respiring organisms, and from the atmosphere.

In somewhat simplified terms, and for the time being ignoring the interchange of carbon dioxide gas with the atmosphere, the dissolution of carbon dioxide in water can be thought of as producing the acid H_2CO_3.

The dissociation constants for this acid are:

$$H_2CO_3 \rightleftharpoons H^+ + HCO_3^- \qquad \log_{10} K_1 = -6.4 \tag{3.4}$$

$$HCO_3^- \rightleftharpoons H^+ + CO_3^{2-} \qquad \log_{10} K_2 = -10.3 \tag{3.5}$$

$$K_1 = \frac{[HCO_3^-][H^+]}{[H_2CO_3]}$$

can be re-written as:

$$\log_{10} K_1 = \log_{10}\frac{[HCO_3^-]}{[H_2CO_3]} + \log_{10}[H^+]$$

or

This is a variant of the Henderson-Hasselbach equation often written as:

$$pH = pK - \log_{10}\frac{[Acid]}{[Salt]}$$

$$pK_1 = -\log_{10}\frac{[HCO_3^-]}{[H_2CO_3]} + pH \tag{3.6}$$

When the pH of the solution is below pK_1 therefore, the main carbonate species is H_2CO_3 and there is little HCO_3^-.

The pH of the solution is at the same time also below pK_2 and there can therefore be only a negligible concentration of CO_3^{2-}.

Below pH 6 almost 100% of the carbonate will be present as H_2CO_3 i.e. the proportion of the carbonate present as H_2CO_3 will be 1. Above the point where the pH = pK_1 the H_2CO_3 is deprotonated and the proportion of the carbonate present in this form drops to zero. A graph can be drawn to show this information with α being the proportion of the carbonate present in a particular form.

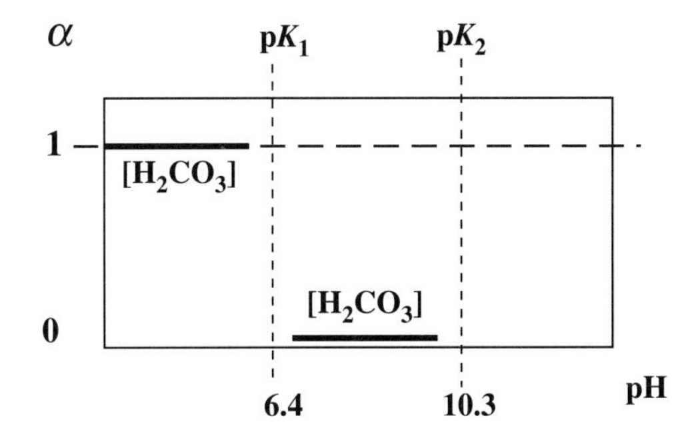

Whilst below pH 6.4 the proportion of the carbonate which is present as HCO_3^- is very low (essentially 0), above that pH it becomes the predominant species and its α rises to 1.

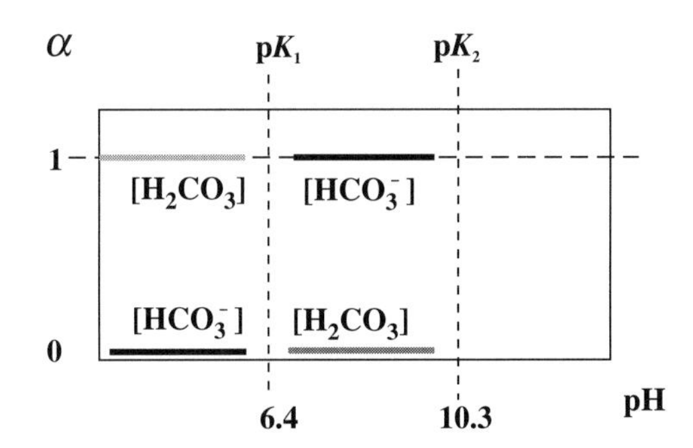

So far, the figure is incomplete as we have not, as yet, identified the crossover point of the HCO_3^- and H_2CO_3 curves. Returning to eqn 3.6, however, we can easily calculate where this point is. Crossover occurs when $[H_2CO_3] = [HCO_3^-]$ i.e. when $pH = pK_1$.

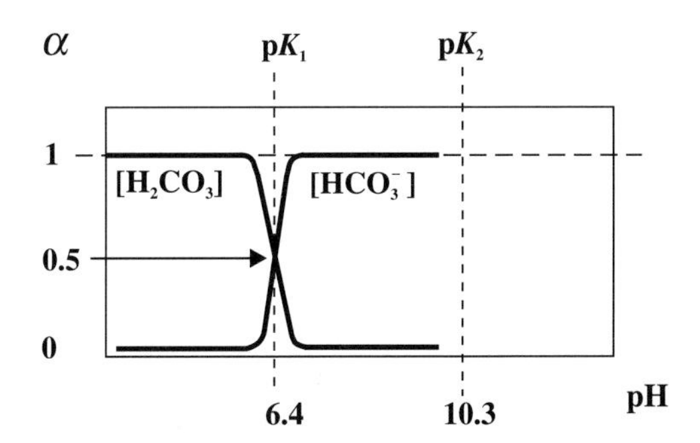

When $[H_2CO_3] = [HCO_3^-]$ the log term becomes zero and the pH therefore equals pK_1. At this point the carbonate is present as 50% H_2CO_3 and 50% HCO_3^-, hence α is 0.5 for both species when $pH = pK_1$.

The same process can be applied to sketch the proportion of the carbonate present in all its possible forms:

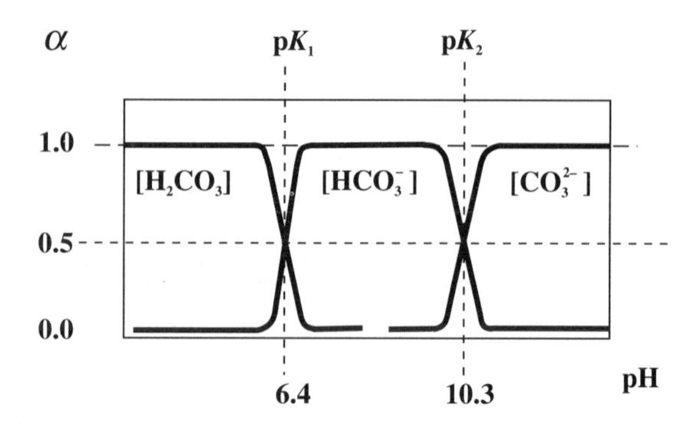

3.5 The solubility of gases

Two dissolved gases, oxygen and carbon dioxide, play major roles in the maintenance of life in aquatic systems. In order to be able to assess the status of a water system it is necessary to understand something about the transport of gases in and out of a body of water. In the simplest examples of gas dissolution, the gas is only solvated by the water and does not react to give a different chemical species. For gas X the equilibrium:

$$X_g \rightleftharpoons X_{aq}$$

can be described by an equilibrium constant, the Henry's Law constant K_H.

This could be based on units of e.g. mol dm^{-3} Pa^{-1} or mol dm^{-3} atm^{-1} depending on the units used to measure the gas pressure.

$$K_H = \frac{[X_{aq}]}{p_X}$$

where $[X_{aq}]$ is the concentration of the gas in aqueous solution and the concentration in the gas phase is measured in terms of its partial pressure p_X.

We can calculate the equilibrium concentration of nitrogen in water at 25 °C using the Henry's Law constant.

Henry's Law constants at 25 °C (mol dm^{-3} atm^{-1})

O_2	1.28×10^{-3}
CO_2	3.38×10^{-2}
CH_4	1.34×10^{-3}
N_2	6.48×10^{-4}

78.08% of air (by volume) is nitrogen and the partial pressure of nitrogen in air, p_{N_2}, is therefore 0.7808 atmosphere.

From Henry's Law

$$[N_{2aq}] = K_H \cdot p_{N_2} = 6.48 \times 10^{-4} \times 0.7808$$

giving a nitrogen concentration of 5.06×10^{-4} mol dm^{-3}.

3.6 Buffering

If the same amount of acid is added to a number of different, but equal sized lakes, they will adopt different pH values. Whilst the pH values of some

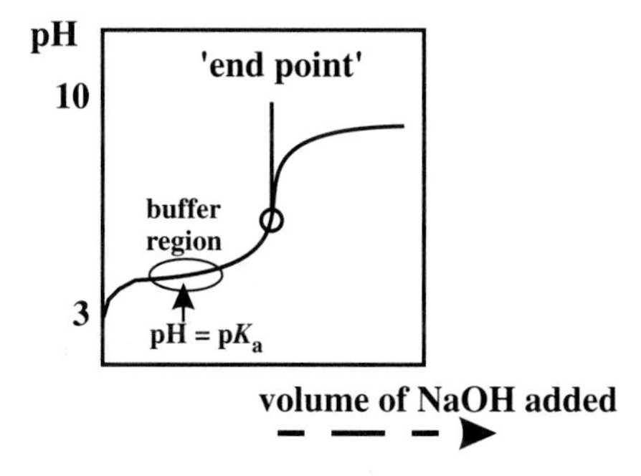

Fig. 3.1 A weak acid being titrated with sodium hydroxide.

natural waters are little affected by the addition of acid others become very acidic. These differences arise from buffering which occurs due to the presence of chemicals dissolved in the water. These can either take up or release protons into solution in response to the added acid (or base). The quantity of acid or base that can be taken up in this way is called the buffering capacity.

A buffer is a compound, or mixture, which helps to maintain the concentration of a species constant.

We have seen earlier that the natural pH of pure water is 7, but it does not however take much to change its pH as pure water has almost no buffering capacity. Rainwater is in equilibrium with atmospheric CO_2 and has a pH of *c.* 5.6. The dissolved carbon dioxide is not a significant buffer and the rain can easily be made more acidic by acidic gases, such as volcanic SO_2 or pollutant gases.

In acid rain, containing sulfurous, sulfuric, and nitric acids, the pH can drop even further. The average pH of rainfall in the eastern US is between 3.9 and 4.5.

A pH buffer is composed of either a weak acid and a salt of that acid, a weak base and a salt of that base, or an acid salt. The buffering effect can be seen in action in a weak acid/NaOH titration curve (Fig. 3.1).

In this titration experiment sodium hydroxide is being gradually added to a solution of a dilute acid, such as acetic acid. Initially the solution is acidic and the NaOH addition has little effect on the solution pH—it is buffered by protons being released from the acid to react with the added hydroxyl ions. This buffered region occurs around pH 4.74, the pH at which pH = pK_a for the acid. Suddenly at the end (or equivalence) point the situation is reached where there is no undissociated acid left to release protons. The solution pH then suddenly jumps to an alkaline value.

The system has so far been described as one in which base has been added to a solution of a weak acid. The reverse is also true. If we were to take away some hydroxyl ions (e.g. by addition of some acid) the pH would be maintained essentially constant in the buffer region around pH 4.74.

Let us now consider the process in reverse and calculate the ratio of sodium acetate to acetic acid which must be mixed to give a solution buffered at pH 5.00. For this example we will work through the calculation trying to use our understanding of the chemistry in solution to identify approximations that can be made.

The equilibrium controlling the pH of the solution is the acid dissociation (deprotonation) of the acetic acid:

$$CH_3COOH \rightleftharpoons H^+ + CH_3COO^-$$

$CH_3COOH = HAc$
$CH_3COO^- = Ac^-$

$$K_a = \frac{[H^+][CH_3COO^-]}{[CH_3COOH]} = 1.8 \times 10^{-5}$$

A solution having a pH of 5.0 contains an $[H^+]$ concentration of 10^{-5} mol dm^{-3} giving:

$$\frac{[CH_3COO^-]}{[CH_3COOH]} = 1.8$$

There are two possible sources of the acetate ion, the acid and the sodium acetate. The acetic acid is only slightly ionized and will therefore make only a small contribution to the acetate ion concentration in solution. On the other hand the sodium acetate is completely dissociated into its component ions and will therefore account for almost all the acetate ions in solution, i.e. $[CH_3COO^-] = C_{salt}$.

The acetic acid concentration [HAc] is essentially unaffected by dissociation and will have the same concentration as the added acid, i.e. $[CH_3COOH] = C_{acid}$.

The ratio of the salt to acid required to give a pH of 5 is therefore:

$$\frac{C_{salt}}{C_{acid}} = 1.8$$

Mixing sodium acetate with acetic acid in the molar ratio 1.8 to 1 will therefore give a pH of approximately 5.0.

There are three important characteristics of a buffer solution. Its pH, the range over which it is effective and how much acid or base can be added to it without the pH changing significantly.

The pH and buffer range

If we take the log form of the K_a expression:

$$K_a = \frac{[Ac^-][H^+]}{[HAc]}$$

$$\log_{10}K_a = \log_{10}[H^+] + \log_{10}\frac{[Ac^-]}{[HAc]}$$

$$pH = pK_a + \log_{10}\frac{[CH_3COO^-]}{[CH_3COOH]}$$

we end up with an equation relating the pH of the solution to the pK_a of the dissolved acid. Looking more closely at the equation it becomes evident that if $[CH_3COO^-] = [CH_3COOH]$ the pH of the solution equals the pK_a.

The logarithmic term becomes zero when $[Ac^-] = [HAc]$

If the ratio $[CH_3COO^-]/[CH_3COOH]$ is greater than one then the pH rises above the pK_a, if the ratio is smaller than one, then the pH drops below the pK_a. One consequence of this is that the pH of the solution depends on the relative concentrations of the acid and its salt, not on their absolute values. 0.001 mol dm^{-3} solutions therefore have approximately the same pH as 1 mol dm^{-3} solutions.

We have also seen that the pH range over which the system buffers is centred around the pK_a. Polyprotic acids such as phosphoric acid, having a number of pK_as, will therefore be able to produce buffer systems operating over a number of pH ranges.

As a general guide, the effective range of a buffer is one pH unit either side of the pK_a.

Buffer capacity

The buffer capacity is a measure of how much acid or base can be added to a solution before its pH changes significantly. In natural water systems this could be the difference between whether the acidity of the water is significantly influenced by the addition of acid rain or is relatively unaffected by it.

In classical terms the buffer capacity would have been defined as the number of moles of strong base required to raise the pH of one litre of solution by one pH unit. By taking a slightly more mathematical approach however, it is possible to get a better understanding of the buffering behaviour of a solution.

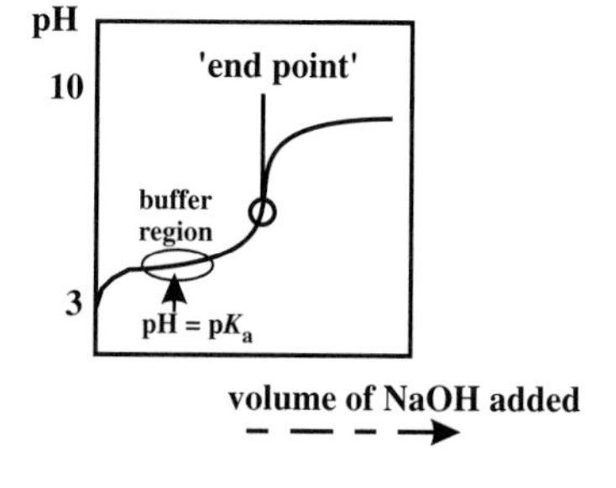

Returning to our weak acid–strong base titration curve (Fig. 3.1), we can see that the buffer capacity is really a measure of the slope of the curve in the buffer region—the poorer the ability of the buffer to maintain a pH, the greater the slope of the curve in this region.

The slope of the curve is called the buffer index β. For finite changes in the concentration of base added to the system this is:

$$\beta = \frac{\Delta C_b}{\Delta \text{pH}}$$

where ΔC_b is the quantity of strong base added and ΔpH is the resulting pH change.

Δ is used mathematically to denote a step change in value.

The way that β changes with pH can be calculated and when presented as a graph highlights the pH range in which a solution is a strong buffer. Figure 3.2 shows such a plot for our acetic acid system. Note how the buffer capacity is maximized when the pH of the solution is at the pK_a of acetic acid (4.74) and how the potency of the buffer around pH 5 depends on the concentration of the solution.

The derivation of the equation is beyond the scope of this book.

We are now in a better position to understand why some water bodies are better able to resist the addition of acid rain than others. Water which drains

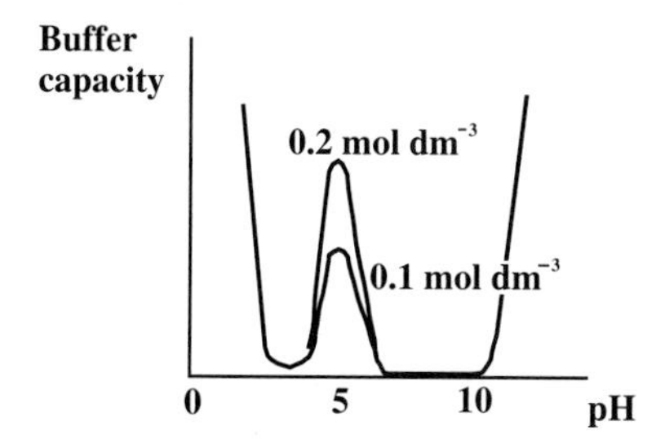

Fig. 3.2 The buffer index for acetic acid solutions.

granite areas contains very little dissolved salts which can act as buffers; the water therefore has little buffering capacity and its pH is therefore readily lowered by acid rain. The water in streams running off chalk areas on the other hand, can effectively counteract the acidification, due to the high concentrations of dissolved carbonates in the water which buffer the pH.

3.7 Case studies

What should the pH of rain be?

When asked what the pH of rain is, most people reply that the natural pH of rain is 7 and that it is only because of pollution that rainwater has become acidic—this is not, however, the case.

The simplest model that we can construct for rainwater is to consider the pH of pure water in equilibrium with atmospheric CO_2.

$$CO_{2_g} + H_2O \rightleftharpoons H_2CO_{3_{aq}}$$

$$H_2CO_{3_{aq}} \rightleftharpoons H^+ + HCO_3^- \qquad K_1 = \frac{[HCO_3^-][H^+]}{[H_2CO_3]}$$

For every H^+ produced, one HCO_3^- is generated; therefore:

Alternative route via charge balance equation:

$$[H^+] = [HCO_3^-] + 2[CO_3^{2-}] + [OH^-]$$
$$[HCO_3^-] = [H^+] - [OH^-] - 2[CO_3^{2-}]$$

For an acidic solution $[OH^-]$ and $[CO_3^{2-}]$ will be negligible therefore:

$$[H^+] = [HCO_3^-]$$

$$[H^+] = [HCO_3^-]$$

$$K_1 = \frac{[H^+][HCO_3^-]}{[H_2CO_3]} = \frac{[H^+]^2}{[H_2CO_3]} \qquad (3.7)$$

It is now necessary to find the concentration of H_2CO_3 in solution; this is determined by the partial pressure of carbon dioxide in the atmosphere and the Henry's law constant:

$$K_H = \frac{[H_2CO_3]}{pCO_2}$$

The $[H_2CO_3]$ in eqn 3.7 can now be replaced by $K_H \cdot pCO_2$. Which then gives:

$$K_1 = \frac{[H^+]^2}{K_H \cdot pCO_2}$$

or

$$[H^+] = \sqrt{K_H K_1 pCO_2}$$

If we assume that the partial pressure of carbon dioxide in the atmosphere is 3.6×10^{-4} atm we can calculate the pH of rainwater ($K_1 = 3.98 \times 10^{-7}$ mol dm^{-3} and $K_H = 0.034$ mol dm^{-3} atm^{-1})

$$\boxed{pH = 5.6}$$

This is the reason that water which is redistilled for laboratory use never has a neutral pH.

—rather more acidic than a pH of 7.

What is the pH of the sea?

In this case study we will build a simple model to describe the chemistry governing the acidity of seawater. The sea is a rather complex mixture, so first we will see how we can simplify its composition.

Seawater contains a very large concentration of dissolved salts but only a few of its components have any significant effect on its pH. Seawater contains very high concentrations of sodium and chloride ions and in solution these have no affinity for each other; i.e. sodium chloride is completely ionized in solution. Most importantly also, Na^+ has no affinity for OH^- ions and Cl^- has no affinity for H^+.

Put another way, sodium hydroxide is a strong base and is therefore completely ionized in solution.

Sodium ions and chloride ions are therefore not going to influence the pH of the sea and can be ignored in our calculation. The pH of the sea can be estimated by considering it to be a simple solution containing calcium ions and carbonate species. As the water is in contact with the atmosphere, the dissolution of gaseous carbon dioxide will be an important factor governing its pH.

Dissolved CO_2 is hydrated and has the formula $CO_2.H_2O$. This is frequently rewritten as H_2CO_3.

An equilibrium calculation, based on the following equations, gives a good estimate:

$$CaCO_3 \rightleftharpoons Ca^{2+} + CO_3^{2-} \quad \log K_{Ca} = -8.3 \qquad (3.8)$$

$$H_2CO_3 \rightleftharpoons H^+ + HCO_3^- \quad \log K_1 = -6.4 \qquad (3.9)$$

$$HCO_3^- \rightleftharpoons H^+ + CO_3^{2-} \quad \log K_2 = -10.3 \qquad (3.10)$$

$$H_2O \rightleftharpoons H^+ + OH^- \quad \log K_W = -14.0 \qquad (3.11)$$

$$pCO_2 + H_2O \rightleftharpoons H_2CO_3 \quad \log K_{CO_2} = -6.5 \qquad (3.12)$$

Equation 3.12 describes the equilibrium of the gaseous carbon dioxide in the atmosphere with dissolved H_2CO_3. p_{CO_2} here employed units of Pa.

The charge balance equation for the system is:

$$2[Ca^{2+}] + [H^+] = 2[CO_3^{2-}] + [HCO_3^-] + [OH^-] \qquad (3.13)$$

Remember to take into account species such as Ca^{2+} contributing two charges.

We want to know the pH and therefore somehow we have to replace all the terms in the equation by either $[H^+]$ terms or ones that we know the value of.

We know the values of equilibrium constants and analytical concentrations.

The way to approach the problem is to look at one term at a time:

[H₂CO₃].

The atmosphere has a CO_2 partial pressure of *c.* 32 Pa ($10^{-3.5}$ atm) and using the Henry's Law constant for eqn 3.12 we obtain:

$32 = 10^{1.51}$
if $32 = 10^x$
$x = \log_{10} 32 = 1.51$

$$[H_2CO_3] = 10^{-6.5} \times 32 = 10^{-5.0} \qquad (3.14)$$

From $K_{CO_2} = \frac{[H_2CO_3]}{p_{CO_2}}$

$$\boxed{[H_2CO_3] = 10^{-5.0} \text{ mol dm}^{-3}}$$

[HCO₃⁻].

Equation 3.9 describes the equilibrium between H_2CO_3 and HCO_3^- and its equilibrium constant K_1, relates the concentration of $[HCO_3^-]$ to those of $[H_2CO_3]$ and $[H^+]$:

$$K_1 = \frac{[H^+][HCO_3^-]}{[H_2CO_3]} = 10^{-6.4}$$

We have just calculated $[H_2CO_3]$ to be 10^{-5} mol dm^{-3}.

therefore

$$[HCO_3^-] = \frac{10^{-(6.4+5.0)}}{[H^+]} = \frac{10^{-11.4}}{[H^+]} \text{ mol dm}^{-3}$$

$[CO_3^{2-}]$.
Now that we know $[HCO_3^-]$ we need an equation relating $[CO_3^{2-}]$ to it. The simplest of these is the acid dissociation expression for K_2:

$$K_2 = \frac{[CO_3^{2-}][H^+]}{[HCO_3^-]}$$

into which should be substituted our equation for $[HCO_3^-]$

$$K_2 = \frac{[CO_3^{2-}][H^+]^2}{10^{-11.4}}$$

giving:

$$[CO_3^{2-}] = \frac{10^{-21.7}}{[H^+]^2} \text{ mol dm}^{-3}$$

$[Ca^{2+}]$.
To derive an expression for $[Ca^{2+}]$ we have to find an equation linking $[Ca^{2+}]$ to terms that are known and/or $[H^+]$. The solubility product equation associated with eqn 3.8 is useful for this as its only terms are K_{Ca} (known), $[CO_3^{2-}]$ (just calculated), and $[Ca^{2+}]$ (wanted).

Note that there is no $[CaCO_3]$ term present as this is a solid.

$$K_{Ca} = [Ca^{2+}][CO_3^{2-}]$$

Eliminating $[CO_3^{2-}]$ gives:

$$K_{Ca} = [Ca^{2+}]\frac{10^{-21.7}}{[H^+]^2}$$

$$[Ca^{2+}] = 10^{13.4}[H^+]^2 \text{ mol dm}^{-3}$$

$[OH^-]$.
To convert $[OH^-]$ to $[H^+]$ just requires the K_W relationship:

$$K_W = [H^+][OH^-]$$

$$[OH^-] = \frac{10^{-14}}{[H^+]} \text{ mol dm}^{-3}$$

We now have all the terms in the charge balance equation in forms which only include known numbers or $[H^+]$. Putting them into the charge balance equation gives:

$$10^{13.7}[H^+]^4 + [H^+]^3 - 10^{-11.4}[H^+] = 10^{-21.4}$$

An equation like this can be difficult to solve. Iterative methods are often used in which a guess is made and this guess is continuously refined until the correct solution is obtained.

This, when solved iteratively, gives $[H^+] = 10^{-8.4}$ mol dm^{-3},

pH = 8.4—not far from the real value.

If the same calculation had been carried out without consideration of the atmospheric contact, the pH would have been approximately 10.

3.8 Problems

3.1 Calculate the pH values of the following 0.1 mol dm^{-3} aqueous solutions: (a) HCl, (b) HF ($K_a = 7.2 \times 10^{-4}$), and (c) ammonia solution ($K_b = 1.8 \times 10^{-5}$).

3.2 Calculate the approximate pH and pOH resulting from the addition of 0.01 cm^3 of concentrated HNO_3 (16 mol dm^{-3}) to 1 litre of water.

3.3 For an aqueous solution containing both sodium carbonate (0.1 mol dm^{-3}) and sodium phosphate (0.2 mol dm^{-3}), give:
(i) the mass balance equation for the phosphate-containing species in solution;
(ii) the charge balance equation for the solution.

3.4 Describe what will happen to the pH of a buffer solution (pH = 4) as it is made increasingly more dilute by the addition of water. What will be the pH at infinite dilution?

3.5 Earlier in this chapter we calculated the pH of rain based on the water being equilibrated with atmospheric carbon dioxide. Now repeat the calculation considering the rainwater to be in equilibrium with the air containing an SO_2 concentration of 5×10^{-9} atm. You can ignore the presence of CO_2.
SO_2 dissolves giving H_2SO_3, with a Henry's Law constant of 1.0. K_1 for H_2SO_3 is 1.7×10^{-2}.

This would not be untypical of the SO_2 concentration over a continental land mass.

3.6 Calculate the solubility of $BaSO_4$ (a) in pure water and (b) in $Ba(NO_3)_2$ solution (0.1 mol dm^{-3}). [K_s for $BaSO_4$ is 1.1×10^{-10}]. Comment on the difference between the two solubilities.

3.7 Describe the chemical compositions of both limestone and granite. Why are natural waters in contact with limestone alkaline?

3.8 Briefly describe two different mechanisms by which acidic gases are deposited from the atmosphere.
How would the pH of a lake be affected by acidic precipitation
(i) in a limestone region?
(ii) in a granite region?

3.9 The titration of a weak base with strong acid is shown in the figure below. Identify the following and describe the chemistry occurring at them
(i) the buffered region (s)
(ii) the equivalence point(s)
How would the reaction profile differ for the titration of a carbonate solution (CO_3^{2-}) with a strong acid?

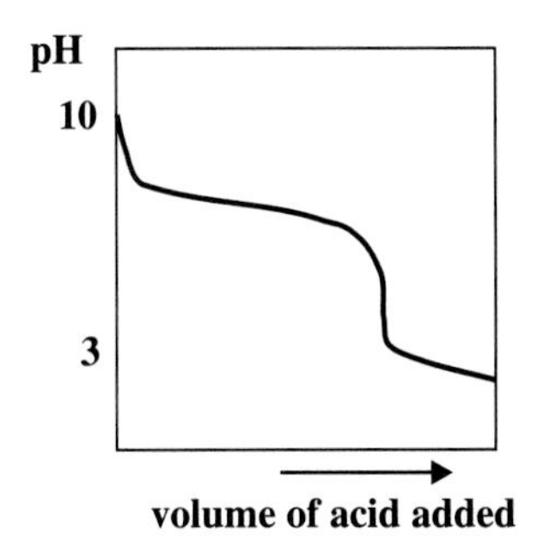

3.10 Calculate the relative concentrations of sodium dihydrogen phosphate (NaH_2PO_4) and phosphoric acid (H_3PO_4) required to give a solution which buffers at pH 2.6. The pK_{a1} for phosphoric acid is 2.1.

3.9 Further reading

Introduction to environmental chemistry. Bunce, N. J. Wuertz Publishing, Winnipeg, 1993. ISBN 0-920063-50-0

Foundations of physical chemistry. Lawrence, C. P., Rodger, A., and Compton, R. G. Oxford University Press, Oxford, 1996. ISBN 0-19-855904

Acid rain; its causes and its effects on inland waters. Mason, B. J. Clarendon Press, Oxford, 1992. ISBN 0-19-8583443

More advanced:

Aquatic Chemistry. Stumm, W. and Morgan, J. J. Wiley Interscience, New York, 1996. ISBN 0-471-51184-6

4 Metal complexes in solution

4.1 Introduction

With modern technology it is possible to find almost every element of the periodic table in a sample of natural water. These samples will differ from each other in the amount of each element or in the chemical form in which it is present. Which metals are present in significant quantities will in the case of groundwaters, streams, rivers, and lakes reflect the geology of the catchment area and the past history of the water. Some of the major factors influencing the concentrations of trace metals in oceanic waters, away from coastal influences, are inputs from bottom sediments and hydrothermal activity; deposition of material from the atmosphere together with removal by chemical adsorption onto surfaces; and precipitation and accumulation by marine organisms.

Not only do the elements which are present in water differ in concentration but the interactions between metal ions and other species in solution mean that each metal can be present in a number of different chemical forms. The chemical or physical form in which an element is present is called its **speciation**. The importance of understanding the **speciation** of an element is that the properties of the element depend on its form. HgS, for example, is a very insoluble species, which would deposit into the sediments of a lake. The Hg^{2+} ion on the other hand is very soluble and will stay dissolved in the water. In terms of toxicity Hg^{2+} is less toxic than the biologically produced methylmercury (CH_3Hg^+). In the case of cadmium, the salinity of the water determines whether the metal is present as Cd^{2+} or $CdCl_4^{2-}$. One of these species is positively charged whilst the other is negatively charged and many properties such as membrane permeability and adsorption onto clay particles will alter as a consequence.

It has also been noted that, for fish, Cd toxicity differs between freshwater and seawater environments. This is believed to be due to the charge differences of the two cadmium species influencing the membrane permeability of the element.

These differences in speciation can arise from a number of different causes. An element's speciation can be governed by the environment in which it is present. It may be altered by the presence of other species in solution such as complexing anions and the acidity or redox properties of the system.

Some of the metal ions, and in particular the Group 1 alkali metals such as sodium and potassium, are believed to be present largely as their hydrated metal ions. The remainder of the metals can occur as a variety of different species by interacting with other compounds in the water to form complexes.

The aquatic chemistry of the metalloid elements of Groups 15 and 16 (As, Sb, Se etc.) is dominated by the formation of oxyanions (such as arsenate, AsO_4^{3-}). Probably as a result of chemical similarities with the nutrient anion species of the non-metals such as phosphate and nitrate, they are sometimes incorporated into the metabolic pathways of the nutrients. Whilst this may result in toxic effects, it may simply result in the release of the element back into the water in a chemically changed form.

You may be pleased to learn that arsenic is stored by fish as the relatively non-toxic compound arsenobetaine ($(CH_3)_3As^+CH_2COO^-$).

As the behaviour of a metal in the environment is so dependent on its chemical form it is important to be able to predict how these changes in form might be brought about. This chapter describes some of the chemistry underlying the formation of metal complexes and outlines some of the techniques which can be employed to provide an insight into the chemical species which might be present in natural and polluted waters.

4.2 Metal complexation

Ligand: electron pair donor = Lewis base.

Metal: electron pair acceptor = Lewis acid

A complex is formed by the donation of electrons from a complexing ligand to a metal. As the ligand is an electron pair donor it is described as being a Lewis base.

The electron pair acceptor (the metal), is by this definition a Lewis acid. The simplest example of a metal complex, and one which is often forgotten, is that of a simple metal ion in aqueous solution. Whilst normally written as, for example, Zn^{2+}, such ions are in fact surrounded by a shell of water molecules. Each water molecule is bonded to the metal by donation of electrons originating from the lone pairs on the oxygen (Fig. 4.1).

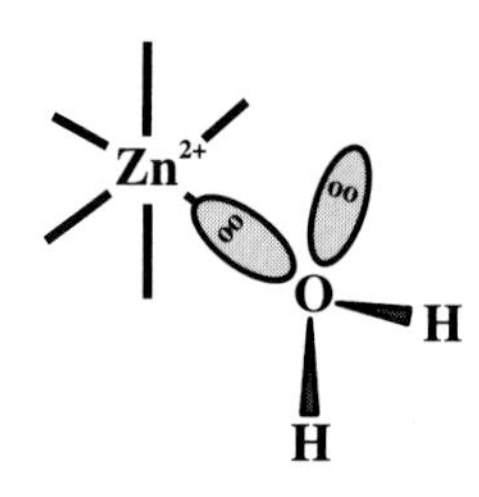

Fig. 4.1 The bonding of a water molecule to a zinc ion in aqueous solution.

A complexation reaction, such as:

$$Zn^{2+} + 4Cl^- \rightleftharpoons ZnCl_4^{2-}$$

is in fact a substitution reaction with chloride ions replacing coordinated water.

$$Zn(H_2O)_6^{2+} + 4Cl^- \rightleftharpoons Zn(H_2O)_2Cl_4^{2-} + 4H_2O$$

4.3 The ligands

Coordinating groups

A wide variety of groups can be involved in coordination, the most important in aquatic systems being:

Halogens. These usually coordinate only as simple anions and rarely do so when they are part of a larger compound.

Oxygen ligands. The most important example of this class of compounds is water. Other examples of functional groups containing oxygen that can take part in complexation include ethers, alcohols, ketones, and carboxylic acids.

Sulfur donors, such as thiols, thioethers, and dithiocarbamate groups.

Nitrogen donors, most commonly in the form of amines.

Denticity

Ligands are distinguished by the number of groups they contain which are capable of binding to a metal. In the complexes so far encountered one coordination position on the metal has been occupied by one ligand molecule. Examples of unidentate ligands, which attach through only one coordinating group, include acetate, CN^- and pyridine. In general these tend to result in the formation of ionic water-soluble complexes.

This happens because whilst a metal ion will at most normally only have a 3+ charge, it will typically be coordinated by about six ligands each having, say, a 1− charge. In such a typical complex the final overall charge of the complex will be at least 3−, making it ionic and normally water soluble.

An example of a cobalt(III) amine complex formed by the coordination of six unidentate ammonia ligands is shown in Fig. 4.2.

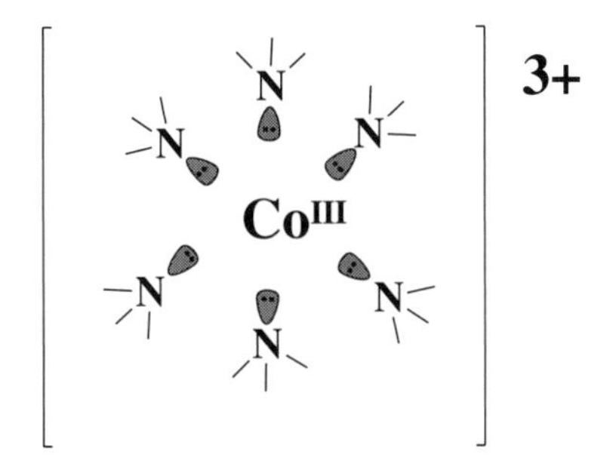

Fig. 4.2 $[Co(NH_3)_6]^{3+}$.

There are many molecules which contain more than one group which is capable of bonding to the metal centre. These are called **polydentate ligands**. A reagent which attaches to a single metal centre in such a way is termed a chelating agent (Fig. 4.3).

Chelate comes from the Greek word chele, the prehensile claw of an arthropod.

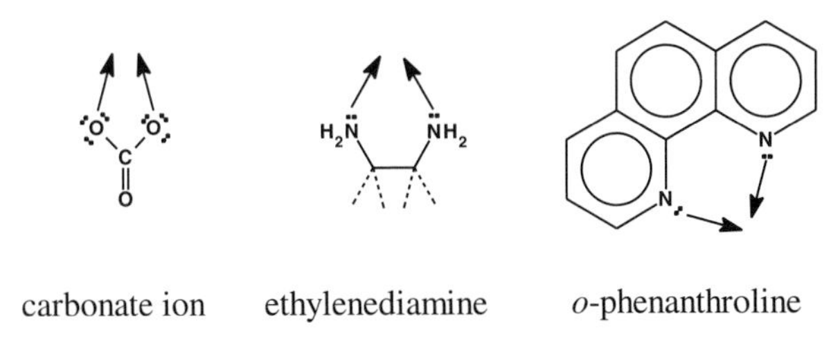

Fig. 4.3 Some examples of bidentate chelating agents.

The number of coordinating groups in the chelating agent and the number of available coordination sites on the metal will govern the stoichiometry of the final compound.

Formation constants

In aqueous solution the formation of a complex between a metal ion M^{n+} and a ligand L, to give the complex ML_n, occurs in steps, each of which is characterized by a formation constant K_i.

Metal ions in solution are surrounded by a coordination sphere of water molecules. Whilst the number of such water molecules varies between metals the most common number is six. The replacement of the first water by a ligand has a formation constant K_1:

You will find that, for brevity, ML^{2+} is used to represent the $M(H_2O)_5L^{2+}$ species. Note how the water has been omitted from the expression for the formation constant as the solvent is water and its concentration is to all intents and purposes unchanged by the position of the equilibrium.

$$M(H_2O)_6^{2+} + L \overset{K_1}{\rightleftharpoons} M(H_2O)_5L^{2+} + H_2O$$

$$K_1 = \frac{[ML^{2+}]}{[M^{2+}][L]}$$

where the square brackets are used to signify the concentration of the species. The magnitude of the formation constant tells us what proportion of the metal is present as ML^{2+} at equilibrium. A high K value indicates that a large proportion of the metal will, at any given instant, be found associated with the ligand and only a small amount will be present as M^{2+}.

The second ligand goes on in the same way:

$$K_2 = \frac{[ML_2^{2+}]}{[ML^{2+}][L]}$$

$$M(H_2O)_5L^{2+} + L \overset{K_2}{\rightleftharpoons} M(H_2O)_4L_2^{2+} + H_2O$$

The overall reaction has resulted in two ligand molecules being added to the metal ion and we could write the following equation to describe it:

$$M(H_2O)_6^{2+} + 2L \rightleftharpoons M(H_2O)_4L_2^{2+} + 2H_2O$$

The symbol employed to describe this overall two-step formation equilibrium is β. We now have two formation constants, one which describes the formation one step at a time (the K values), the other combining all steps together (β). In general terms these can be written:

$$K_i = \frac{[ML_i]}{[ML_{i-1}][L]}$$

$$\beta_i = \frac{[ML_i]}{[M][L]^i}$$

These two are interrelated, β is the product of all the K values used in the formation of the complex:

$$\beta_i = K_1 K_2 \ldots K_i$$

The formation of a complex consisting of six ligands binding to a central metal atom can therefore be described in terms of six K values. Each of these describes the addition of one extra ligand at a time. The overall formation constant for our six ligand case would therefore be given by

$$\beta_6 = K_1K_2K_3K_4K_5K_6$$

4.4 Factors influencing the stability of metal complexes

The nature of bonding

There are two components to any complexation reaction, the metal and the complexing agent, both contribute to the formation of the complex and both have their own characteristics. Only if these characteristics match is a strong bond formed between the two.

Not that different to humans really!

Natural waters are complex systems containing many metals and an even greater number of potential ligands. Predicting which metal will be bound strongly by which ligand requires an understanding of factors controlling the bond formation. From a consideration of the natures of the participating metals and ligands it is possible to break these down into two extreme cases. Those metals which are small in size, have a high charge, and tend to participate in ionic bonding are called 'hard' metals (Table 4.1). At the other end of the scale are the 'soft' metals which are generally larger in size, have a lower positive charge, and tend to participate in covalent bonding.

In a similar manner the ligands can be divided into two extremes describing their properties as electron donors, giving us 'hard' and 'soft' bases (Table 4.2).

Once the metals and ligands have been split into 'hard' and 'soft' the prediction of complex stability is based on the preference of 'hard' acids for 'hard' bases and 'soft' acids for 'soft' bases. This is the basis of what is called Hard Soft Acid Base Theory. Table 4.3 gives some examples of particularly favoured bonding combinations predicted by the HSAB approach.

The wide range of binding strengths that can be encountered can be seen from the formation constants for a number of metal complexes involving the amino acid cysteine.

cysteine = $HS{-}CH_2CH(NH_2)COOH$

Table 4.1 Metals categorized as hard or soft acids

Property	Hard	Soft	Intermediate
Polarizability	Low	High	
Electropositivity	High	Low	
+ve charge/oxidn state	Large	Small	
Size	Small	Large	
Bonding	Ionic/electrostatic	Covalent	
Outer electrons on donor	Few and not easily excited	Several and easily excited	
Examples	K^+, Li^+, Na^+, Ba^{2+}, Ca^{2+}, Mg^{2+}, Al^{3+}	Ag^+, Au^+, Tl^+ Hg^+, Hg^{2+}, Cu^+, Cd^{2+}	Fe^{2+}, Co^{2+}, Ni^{2+}, Cu^{2+}, Zn^{2+}, Pb^{2+}.

Table 4.2 Donor groups categorized as hard and soft bases

Property	Hard	Soft
Polarizability	Low	High
Electropositivity	High	Low
+ve charge/oxidn state	Large	Small
Size	Small	Large
Bonding	Ionic/electrostatic	Covalent
Available empty orbitals or donor atom	High energy /inaccessible	Low lying /accessible
Examples	H_2O, F^-, Cl^-, ClO_4^-, RNH_2, OH^-, CO_3^{2-}, NO_3^-, SO_4^{2-}	RSH, I^-, SCN^-, CN^-, R_3As

In between these two extremes there are many metals and ligands which exhibit intermediate behaviour.

Table 4.3 Predicted successful metal–ligand combinations

Metal	Ligand Groups
K^+	singly charged O donors or neutral O ligands
Ca^{2+}, Mg^{2+}, Mn^{2+}	carboxylate, phosphate, N donors
Fe^{2+}	-SH, $-NH_2$ > carboxylates
Fe^{3+}, Co^{3+}	carboxylate, $-NH_2$
Cu^+, Ag^+, Hg^{2+},	-SH
Cu^{2+}	amines >> carboxylates

Table 4.4 The stability of metal–cysteine complexes

	Mg^{2+}	Mn^{2+}	Cu^+
K_1:	$<10^4$	$10^{4.1}$	$10^{19.2}$

The particular stability of chelates

ammonia = NH_3
en = $NH_2(CH_2)_2NH_2$
dien = $NH_2(CH_2)_2NH(CH_2)_2NH_2$

The three ligands ammonia, 1,2-diaminoethane (ethylenediamine (en)), and diethylenetriamine (dien), are chemically very similar, each consisting of a number of potentially coordinating amine groups joined by a number of short hydrocarbon chains. The formation of a nickel complex with six ammonia ligands would at first sight seem to be very similar to the formation of $Ni(en)_3$ or $Ni(dien)_2$. In each case the nickel is coordinated to 6 nitrogen atoms. When the formation constants for these three complexes are measured however they are very different from each other. To understand why, we should compare the formation of complexes containing 6 coordinated nitrogen atoms.

In terms of the number of coordinating amine groups:
3 en = $6NH_3$
2 dien = $6NH_3$

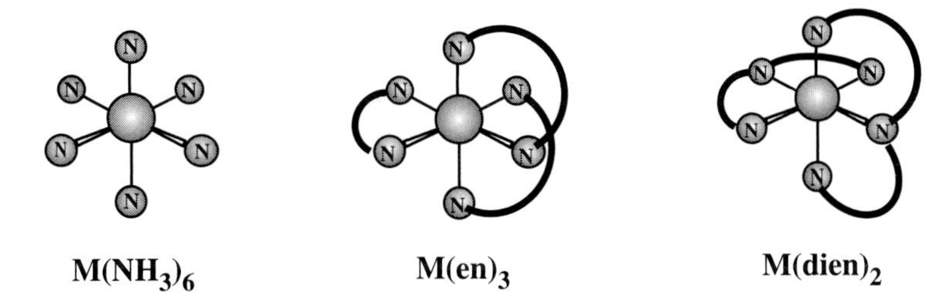

We therefore require β_6 for the ammonia complex, β_3 for the en chelate complex, and β_2 for the dien complex (Table 4.5). The en and dien chelate complexes are significantly more stable than the ammonia complex and there is clearly therefore an advantage to be gained in terms of stability from the formation of a chelate rather than a complex. This is called the **chelate effect** and results from the change in entropy which arises from the replacement of six water molecules by three molecules of ethylenediamine.

When six water molecules are replaced by three en molecules an additional three particles (ammonia molecules) are released into solution, increasing the entropy.

Table 4.5 A comparison of six-coordinate nickel–amine complexes

Ligand	Complex	Formation constant
NH_3	$Ni(NH_3)_6^{2+}$	$\beta_6 = 1 \times 10^8$
en	$Ni(en)_3^{2+}$	$\beta_3 = 4 \times 10^{18}$
dien	$Ni(dien)_2^{2+}$	$\beta_2 = 8 \times 10^{18}$

An additional factor which governs the stability of a chelate complex is the size of the ring which is formed between the chelating agent and the metal. tn is similar to en in most respects, differing only by having one more CH_2 group between the coordinating nitrogens (Fig. 4.4).

tn $= NH_2CH_2CH_2CH_2NH_2$ forms six membered rings

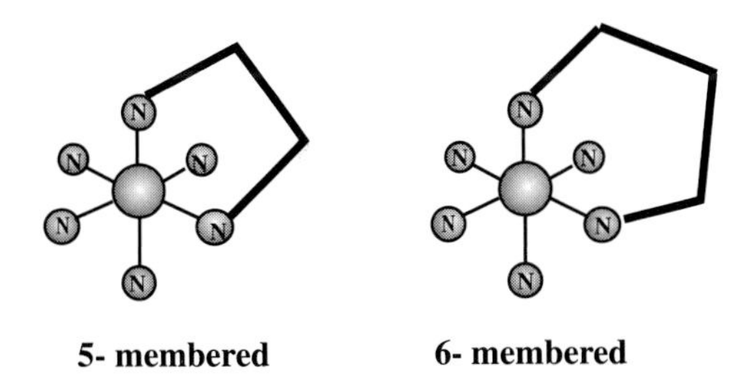

Fig. 4.4 Structural features of five- and six membered ring metal chelates.

The differences between the stabilities of the chelates formed using agents giving five- and six-membered rings is reflected in the stability constants of the resulting complexes:

$$Ni^{2+}\text{–en complexes } \beta_3 = 4 \times 10^{18}$$
$$Ni^{2+}\text{–tn complexes } \beta_3 = 10^{12}$$

In most cases the five-membered ring systems are found to be the most stable and this is believed to be due to the strain induced on the molecule and poor orbital overlap in the larger ring system.

4.5 Describing what happens in natural waters

Having now established the chemical basis of metal complexation it is necessary to apply our understanding of the chemistry of simple systems to a much more complex environmental system.

One of the first things to note is that there is no real reason to expect a reaction to go to completion. Take as an example the reaction between the Hg^{2+} ion and the chloride ion. This is a multi-stage process, the first step being:

$$Hg^{2+} + Cl^- \rightleftharpoons HgCl^+$$

The addition of one more chloride ion gives:

$$HgCl^+ + Cl^- \rightleftharpoons HgCl_2$$

and this reaction goes on yielding $HgCl_3^-$ and $HgCl_4^{2-}$.

OH^- is produced or consumed in response to pH changes. At high pH the concentration of OH^- will be high and it will therefore able to compete effectively against the Cl^- ion for coordination positions on the metal.

So far we have five mercury species, but the story does not stop there. We have not as yet taken into account the presence of the hydroxyl ion. Bringing in the hydroxyl ion yields a number of extra mercury species such as:

$$HgCl^+ + OH^- \rightleftharpoons HgOHCl$$

Even in this simple mercury solution we have already found that a large number of equilibrium reactions are occurring. If we are to be able to understand what is going on we need a good way of describing the system. In order to assess the environmental impact of a trace metal in a water body, predictions have to be made as to which species are present in solution. Even with the power of modern computation, such calculations can take a very long time for all but the simplest systems, and the results generated are not necessarily easy to interpret. A graphical representation normally helps the investigator to understand what is going on.

As they say: 'a picture is worth a thousand words'

Three types of diagram are to be introduced:

(1) concentration ratio diagram;

(2) log(concentration) diagram;

(3) stability field diagram

The concentration ratio diagram

This type of diagram is a useful indicator of major species and trends, but as it considers each equilibrium in isolation, it can easily be over-interpreted. Do not therefore attempt to gather too much information from the minor species.

As its name implies, this diagram shows the ratio of the concentration of a particular metal species to the concentration of free metal ion; this is often expressed as a function of a variable such as pH. The diagram is easy to construct as it is built up from the individual equilibrium equations considered out of context of the other equilibria.

Let us construct a diagram to show how mercury speciation changes with pH. First, a simple complexation step which is independent of pH.

$$Hg^{2+} + 4Cl^- \rightleftharpoons HgCl_4^{2-} \quad \log_{10}K = 15.2$$

$$K = \frac{[HgCl_4^{2-}]}{[Hg^{2+}][Cl^-]^4}$$

To obtain the concentration ratio, rearrange the equilibrium equation

$$\frac{[HgCl_4^{2-}]}{[Hg^{2+}]} = 10^{15.2}[Cl^-]^4$$

and convert it into logarithmic form

$$\log_{10}\frac{[HgCl_4^{2-}]}{[Hg^{2+}]} = 15.2 + 4\log_{10}[Cl^-]$$

For seawater, we can assume $[Cl^-] = 0.6$ mol dm^{-3}, therefore:

$$\boxed{\log_{10}\frac{[HgCl_4^{2-}]}{[Hg^{2+}]} = 14.31}$$

This can be represented graphically as a line on a graph of $\log_{10}\{[HgCl_4^{2-}]/[Hg^{2+}]\}$ versus pH (Fig. 4.5).

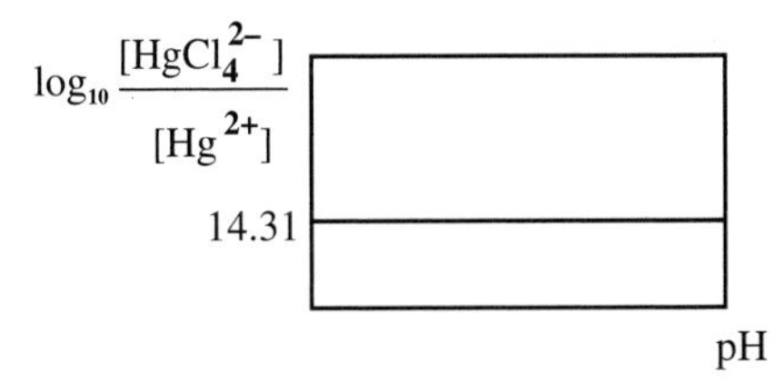

As the reaction is independent of pH the line is parallel to the pH axis.

Fig. 4.5 The $HgCl_4^{2-}/Hg^{2+}$ concentration ratio line.

The hydrolysis of mercury(II) on the other hand does depend on pH:

$$Hg^{2+} + 2H_2O \rightleftharpoons Hg(OH)_2 + 2H^+ \quad \log_{10}K = -6$$

For which:

$$K = \frac{[Hg(OH)_2][H^+]^2}{[Hg^{2+}]}$$

$$\boxed{\log_{10}\frac{[Hg(OH)_2]}{[Hg^{2+}]} = \log_{10}K + 2\text{pH} = -6 + 2\text{pH}}$$

Continuing with other equilibrium equations, the log (concentration ratio)/pH diagram for mercury(II) in seawater can be constructed (Fig. 4.6):

The highest concentration species is at the top of the diagram

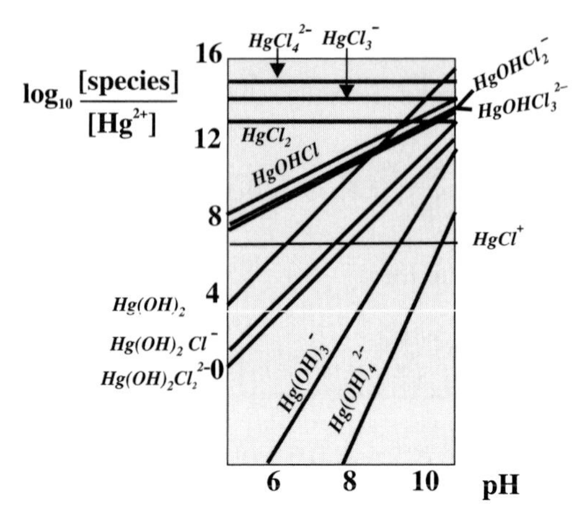

Fig. 4.6 A log(concentration ratio) diagram for mercury in seawater.

The main conclusions that can be drawn from such a diagram are the identification of the relative importances of the species. By looking to the top of the diagram it can be seen that up to a pH of *c.* 11, the most important species is $HgCl_4^{2-}$. Its concentration relative to Hg^{2+} is constant with pH. At very high pH (above 11), the top line becomes that of $Hg(OH)_2$. Less significance can be drawn from the absolute concentrations of the other species in the diagram, but it is possible to identify the relative importances of the minor species. In this particular example

$$[HgCl_4^{2-}] > [HgCl_3^-] > [HgCl_2]$$

The log(concentration) diagram

Our second type of diagram is potentially the most informative. As its name implies, it is nothing more than a plot of species concentration versus some variable such as pH.

We are again going to look at the Hg/Cl^-/pH system but as we can not deal with the chlorinity and pH together we are going to fix [Cl^-] at 10^{-3} mol dm^{-3} (a freshwater). The total concentration of mercury in our system will be 10^{-7} mol dm^{-3}.

The diagram for our system is shown in Fig. 4.7. The dominant species in this system, which contains only a very low chloride ion concentration, are $HgCl_2$ and $Hg(OH)_2$.

Most freshwater systems have pHs in the range 6 to 9—the most complex region of the diagram and a region in which a small change in pH can lead to quite severe changes in speciation.

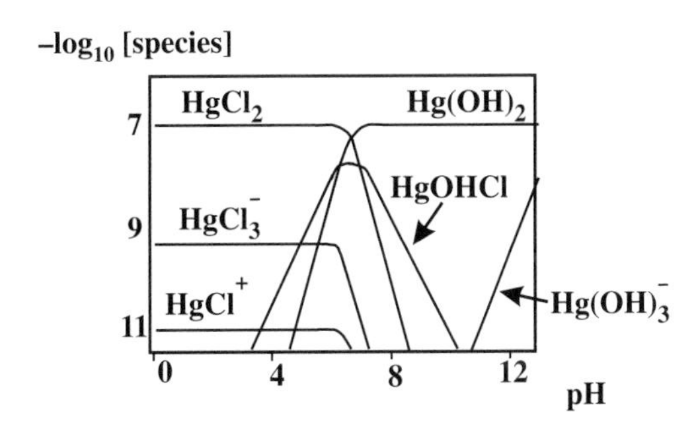

Fig. 4.7 A log(concentration) diagram showing mercury speciation under low chlorinity conditions.

For the construction of Fig. 4.6 a high concentration of chloride ion had been assumed to represent seawater. This resulted in the dominant species in seawater being identified as $HgCl_4^{2-}$. In this example the chloride ion concentration is much less and the chloro-species of mercury are therefore less significant.

Notice the positions of the various types of species, hydroxy (to the right) and chloro (to the left), not altogether surprising as the system can be considered to be a fight between these two ligands for the metal centre.

If we now look at the corresponding diagram for a more saline situation ($[Cl^-] = 0.1$ mol dm^{-3}), corresponding maybe to a part of an estuary, the chloro-species region is seen to extend further over to the right (Fig. 4.8).

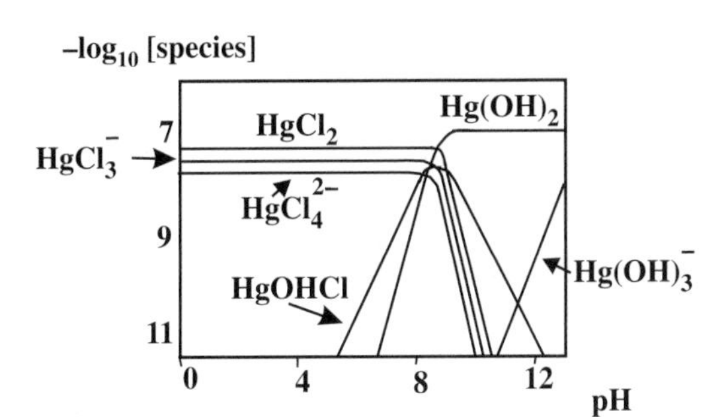

Fig. 4.8 A log(concentration) diagram showing mercury speciation under moderately high chlorinity conditions (estuarine).

Almost as if deliberately however, the transition region is now again on top of the most common pH for saline waters (pH 8.2)!

Stability field (predominant species) diagrams

Whereas the previous two types of diagram have been limited to one changing parameter, such as pH, this diagram allows us to demonstrate the changes of speciation as a function of two variables at a time. In Fig. 4.9 the two variable chosen are pH and pCl.

$pCl = -\log_{10}[Cl^-]$

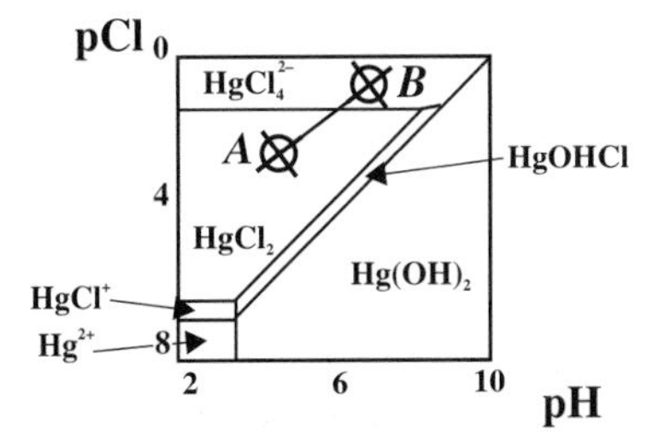

Fig. 4.9 The pH–pCl stability field diagram for mercury species.

As with most real life farms, weeds are to be found in the crop fields, but hopefully they do not dominate the field!

The diagram shows the boundaries within which a particular species predominates—it is rather like an aerial view of a farm, showing the major crops in each field.

Point A is within the $HgCl_2$ field signifying that the highest concentration mercury species under these conditions is $HgCl_2$. If the pH and pCl conditions are changed by walking towards point B, the fence between the fields is encountered. This is the boundary between the $HgCl_2$ field and the $HgCl_4^{2-}$ field. At this boundary $[HgCl_2] = [HgCl_4^{2-}]$.

The next section, dealing with redox speciation, will show you how to construct this type of diagram.

In our simple mercury system we can still see that the most abundant species are the chloro-species under conditions of high chorinity (low pCl) and low pH. In freshwater situations, in which the pCl is high (low $[Cl^-]$), we have $Hg(OH)_2$ and HgOHCl.

4.6 The major complexes of trace elements

For most elements there is believed to be little difference in the nature of dissolved trace element speciation between freshwater and seawater (Table 4.6). The major differences occur with those elements which form strong (halo) complexes or ion pairs (Table 4.7).

Table 4.6 Elements predicted to have similar speciation in freshwater and seawater

Element	Major species
Na	Na^+
C	HCO_3^-
Cl	Cl^-
Cr	$Cr(OH)_3^0$, CrO_4^{2-}
As	$HAs^VO_4^{2-}$, $H_2As^VO_4^-$
I	IO_3^-, I^-
Pb	$PbCO_3^0$, $Pb(CO_3)_2^{2-}$

Table 4.7 Elements predicted to differ in their speciation in freshwater and seawater environments

Freshwater	Both	Seawater
	Mn^{2+}	$MnCl^+$
Ag^+		$AgCl_2^-$
Cd^{2+}, $CdOH^+$		$CdCl_2$
$Au(OH)_3$		$AuCl_2^-$
$Hg(OH)_2$, HgOHCl		$HgCl_4^{2-}$
HPO_4^{2-}		$MgPO_4^-$
SO_4^{2-}		$NaSO_4^-$
F^-		MgF^+

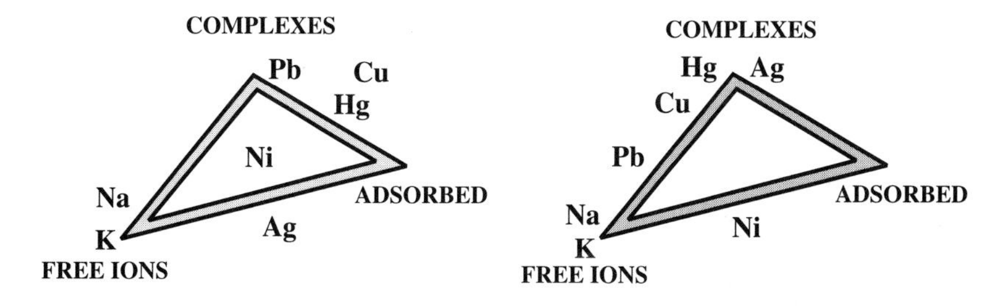

Fig. 4.10 Distribution of metals between free ions, complexes, and species adsorbed onto surfaces.

The changes in the types of species which occur for a variety of trace metals in seawater and freshwater, are shown in Fig. 4.10. Notice how in the freshwater model the elements are divided between the three categories. In the seawater case, however, the high ion concentrations in the water result in the blocking of adsorption sites on particulate material and strong complexation of the metals. The overall result is the comparatively low importance of the adsorbed species.

4.7 Case studies

Man's release of complexants into the environment

Modern-day washing powders and liquid detergents are complex chemical mixes having a number of components which are added to assist the detergent in its action. For many years phosphates (normally polyphosphates) have been a major component of washing products, being added to react with magnesium and calcium ions and to prevent them forming precipitates with soaps or impeding the action of detergents.

For a number of reasons, not least of which has been concern over the release of phosphate into the aquatic environment, in some formulations these phosphates have been replaced by chelating agents such as nitrilotriacetic acid ($N(CH_2COOH)_3$, NTA) and ethylenediaminetetraacetic acid ($(CH_2N(CH_2COOH)_2)_2$, EDTA).

NTA is a tridentate ligand chelating through two of its three carboxylate groups and also through the nitrogen. EDTA is a particularly flexible hexadentate ligand.

Such compounds are not totally benign. Whereas the concern over the presence of phosphates was largely focused on eutrophication, the concern over the release of chelating agents into the environment is based on their potential to remobilize heavy metals such as lead, cadmium, and mercury from the sediments into the water column.

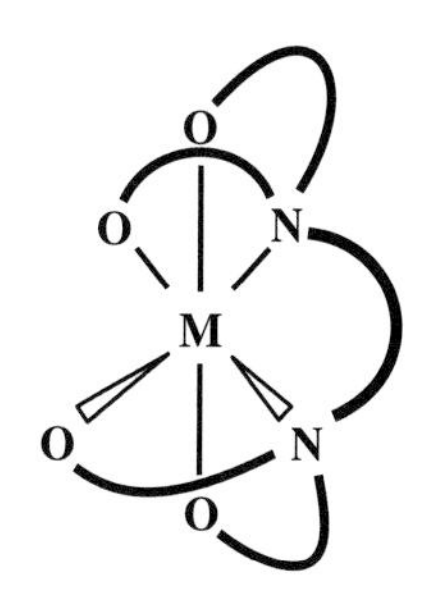

Typical metal–EDTA complex

Pollutant metals can be considered to exert their least environmental impact when immobilized within the bottom sediments of polluted lakes and rivers. The sudden release of the heavy metals back into the water column would be expected to cause a rapid alteration of the ecological balance of the water body by killing certain levels of the food chain. A slow release of metals on the

Biomagnification is the process by which the concentration of an element is increased when it is accumulated by an organism. A typical aquatic plant, for example, may contain a concentration of copper 100 times higher than the concentration in the water. A mollusc feeding on that plant may contain a concentration 10 times higher than that in the plant on which it feeds.

other hand might have less acute impact, but by biomagnification through the food chain might result in the poisoning of higher organisms (of which we are one). By carrying out some simple equilibrium calculations we can assess the potential magnitude of the problem.

The example we are going to investigate is the potential solubilization of $PbCO_3$ by NTA. $PbCO_3$ is one of the most common insoluble lead species, being formed by the action of atmospheric CO_2 and dissolved carbonate ions of lead and its salts. In water its breakdown to yield the Pb^{2+} ion is very limited and can be described by the equilibrium equation:

$$PbCO_3 \rightleftharpoons Pb^{2+} + CO_3^{2-} \qquad K_s = 10^{-13.1}$$

NTA is a tribasic acid. Using H_3NTA to represent the fully protonated acid:

By inspection of the pK values we can see that between pH 2.95 and pH 10.28 the major NTA species will be $HNTA^{2-}$.

$$H_3NTA \rightleftharpoons H^+ + H_2NTA^- \qquad pK_{a1} = 1.66$$

$$H_2NTA^- \rightleftharpoons H^+ + HNTA^{2-} \qquad pK_{a2} = 2.95$$

$$HNTA^{2-} \rightleftharpoons H^+ + NTA^{3-} \qquad pK_{a3} = 10.28$$

The NTA speciation changes with pH can be seen from a plot of the fraction (α) of the NTA present as a particular species:

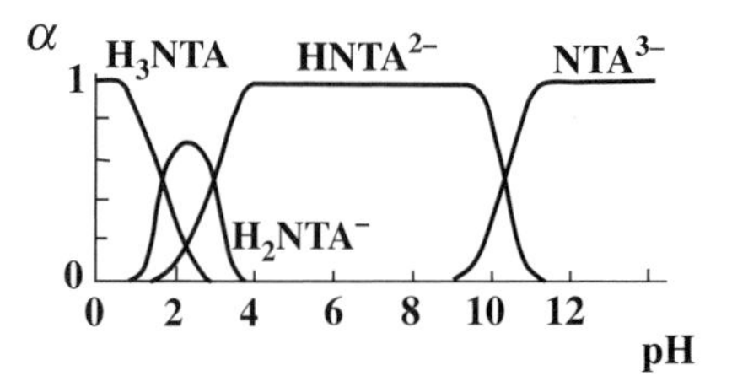

As most natural waters have pH values in the range 4–9, the major form of NTA will be $HNTA^{2-}$.

NTA forms a strong complex with Pb^{2+} ions:

$$Pb^{2+} + NTA^{3-} \rightleftharpoons PbNTA^- \qquad K_{Pb} = 10^{12.6}$$

Taking into account the forms in which the lead and NTA are present, the overall reaction between NTA and $PbCO_3$ can be described by the equation:

$$PbCO_3 + HNTA^{2-} \rightleftharpoons PbNTA^- + HCO_3^- \qquad K_x = 10^{-0.48}$$

How did we obtain a value for the equilibrium constant of the reaction? The first thing to notice is that if we add together the following equations, and then cancel out the species which are to be found on opposite sides of the resulting equation, we are left with the equation we want.

$$PbCO_3 \rightleftharpoons Pb^{2+} + CO_3^{2-} \qquad K_s = 10^{-13.1}$$
$$Pb^{2+} + NTA^{3-} \rightleftharpoons PbNTA^- \qquad K_{Pb} = 10^{12.6}$$
$$HNTA^{2-} \rightleftharpoons H^+ + NTA^{3-} \qquad K_N = 10^{-10.28}$$
$$CO_3^{2-} + H^+ \rightleftharpoons HCO_3^- \qquad K_a = 10^{10.3}$$

To obtain the equilibrium constant we must multiply the component K values.

$$K_s \cdot K_{Pb} \cdot K_N \cdot K_a$$
$$= [Pb^{2+}][CO_3^{2-}] \cdot \frac{[PbNTA^-]}{[Pb^{2+}][NTA^{3-}]} \cdot \frac{[NTA^{3-}][H^+]}{[HNTA^{2-}]} \cdot \frac{[HCO_3^-]}{[H^+][CO_3^{2-}]}$$
$$= \frac{[PbNTA^-][HCO_3^-]}{[HNTA^{2-}]}$$
$$= K_X$$

Formation constants etc. are often given as their $\log_{10}K$ values and these can simply be added to obtain the $\log_{10}K_X$.

The formation constant of $10^{-0.48}$ (0.331) indicates that significant dissolution of lead carbonate can occur in the presence of NTA.

It is interesting at this point to consider the processes involved in the release of metal into solution flowing through a $PbCO_3$-coated lead pipe or washing over sediments containing lead immobilized as lead carbonate. Whilst we might be able to assume that the NTA solution very rapidly reaches a localized equilibrium with the $PbCO_3$ surface, as soon as it has done so, the solution will have moved away from the surface to be replaced by a fresh solution which in turn removes some more lead. It would not therefore be necessary for the solubilization equilibrium constant to be particularly large for significant lead dissolution to occur.

In natural systems, in which many metal ions and potential complexants are present, competition between metals for a complexation site can be crucial. Calcium complexes with NTA are weaker than those formed with lead:

$$Ca^{2+} + NTA^{3-} \rightleftharpoons CaNTA^- \qquad K = 10^{7.6}$$

but in most natural waters the concentration of available dissolved calcium will generally be much higher than the concentration of lead available from $PbCO_3$ dissolution. This strength of numbers permits the calcium to compete effectively against the lead for the available NTA complexation sites, thereby suppressing the lead dissolution.

The accumulation of heavy metals by aquatic organisms

It has long been recognised that some marine organisms, and in particular shellfish, were to be valued nutritionally for their high trace metal content. In exceptional circumstances, normally as a result of localized pollution, their ability to concentrate metals such as copper, zinc, cadmium and mercury has lead to instances of poisoning. The ability of the mussel (*Mytilus* sp.) to accumulate trace metals has been exploited as a means of monitoring the levels of water pollution around the world.

The mussel accumulates metals such as Zn, Cd and Pb over a long period of time thereby smoothing out any fluctuations in metal concentration which might occur in the water; in other words it acts as an integrator. The resulting high levels of metal in the mussel are also much easier to measure than the low levels in the water.

How do marine shellfish accumulate so much metal? The answer lies in the discovery in 1975 of a low molecular weight protein in the common marine limpet, *Patella vulgata*. This protein bound metals strongly and was similar to a metal-binding protein previously found in the kidneys of horses; the protein was called metallothionein. Subsequent investigations led to the discovery of metallothioneins in a number of other marine organisms.

Metallothioneins have a number of interesting properties but the most important are their molecular weight (~6000–7000 Da), and very high metal content, and that ~30% of the protein is made up of the SH amino acid cysteine.

The relative affinities of metals for metallothionein lie in the sequence Hg > Cu > Cd > Zn > Ni, Co. Soft metals have a strong affinity for the soft base SH groups on metallothionein.

Cadmium, copper and zinc are commonly associated with the protein and there is a fixed relationship between the metal content of the protein and the number of cysteine groups in the protein. At pH 7 metallothionein forms very strong metal complexes, with stability constants for the reaction between the metal and metallothionein being ~10^{18} for copper, 10^{15} for cadmium and 10^{12} for zinc.

In HSAB terms the thiol group is soft and metallothionein has an affinity for soft metals.

It therefore appears that the reason that many marine organisms accumulate large quantities of heavy metals is that they are able to produce an effective natural chelating agent in the form of metallothionein. Whilst this protein is principally induced by zinc, and its main function may well be to ensure that zinc is always available for normal metabolic processes, its presence leads to the bioaccumulation of the more toxic elements which have stronger affinities for the strongly chelating thiol groups on metallothionein.

4.8 Problems

1. Draw the structure of a complex formed by the reaction of Fe^{3+} with o-phenanthroline.
2. Using the structure shown in Figure 2.7 identify the main donor groups present in typical humic material. Predict which of the following metals might be expected to have a strong affinity for humic material: Zn^{2+}, Al^{3+}, Cd^{2+}, Ag^{+}, Ca^{2+}, Mg^{2+}.
3. The diagram below illustrates the changes in cadmium speciation which occur as the chloride ion concentration is altered. Alpha (α) is the proportion of the cadmium present in a particular form.

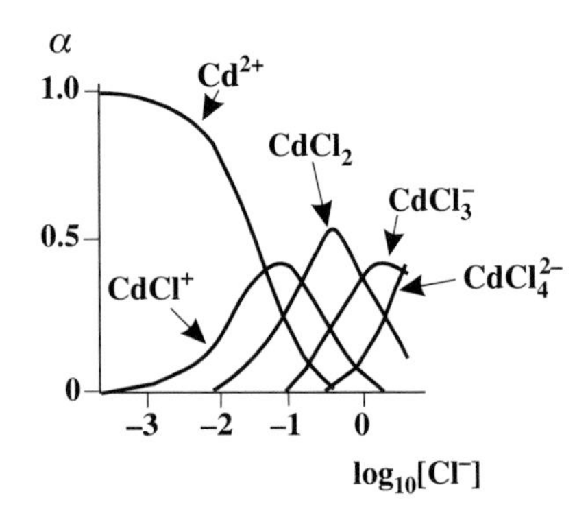

A factory waste containing dissolved cadmium is discharged into a river.

(i) Outline the changes which will occur in cadmium speciation as the dissolved cadmium moves down the river and through an estuary into the sea.

(ii) In many systems metals are often associated with the surface of particles which are suspended in the water. The basis of this association can be a relatively weak ionic affinity of the positively charged metal ion for a negatively charged particle surface. Suggest how the association of the cadmium with these particles might be influenced by their movement through the estuary system.

4.9 Further reading

Metal complexation

Basic inorganic chemistry. Cotton, F. A., Wilkinson, G., and Gaus, P. L. Wiley, Chichester, 1995. ISBN 0471 599743

Physical inorganic chemistry. Kettle, S. F. A. Spektrum, Oxford, 1996. ISBN 0-7167-4554-2

Essentials of inorganic chemistry I. Mingos, D. M. P., Oxford Chemistry Primers. Oxford University Press, Oxford, 1995. ISBN 0-19-855 848-1

Inorganic chemistry. Shriver, D. F., Atkins, P. W., and Langford, C. H. 2nd Edition, Oxford University Press, Oxford, 1994. ISBN 0198 55396 X

The diagrams

Aquatic chemistry. Stumm, W. and Morgan, J. J. Wiley Interscience, New York, 1996. ISBN 0-471-51184-6

A classic fairly advanced text.

Ionic equilibrium, a mathematical approach. Butler, J. N. Addison Wesley, Massachusetts, 1964

Advanced and not easy but clear and well presented.

5 Oxidation and reduction

5.1 Introduction

When the muddy sediments at the bottom of a river or lake are disturbed, two things are often noticed. First, there is a smell of rotten eggs, indicative of the generation of hydrogen sulfide. Secondly, the colour of the sediment may change from a red–brown at the surface to a dark brown–black below. Both of these observations are linked to a lack of oxygen. The hydrogen sulfide will have come from the metabolism of sulfate by sulfate-reducing bacteria, the dark coloration is due to the presence of metal sulfides in the sediment which have been formed from this hydrogen sulfide.

A large proportion of the main group/transition and 'heavy' metal sulfides are very insoluble and therefore find a natural home in the sediments.

In purely chemical terms the absence of oxygen in these sediments has resulted in the reduction of sulfur(VI) to sulfur(–II); that the reduction has been carried out by bacteria does not necessarily alter the chemical result.

Where air is present in the surface of the sediments the sediments are described as being *oxic* (in the presence of free oxygen), in the region where the air is essentially absent, the sediments are *anoxic* (without free oxygen).

It is not only the sediments which can become devoid of oxygen. The water in lakes, and even the bottom waters of fjords, can reach this state if the oxygen is consumed by organisms and chemical processes faster than it can be replaced. This normally occurs when the water is prevented from mixing with oxygenated water by density gradients in the water column and/or constraint by geological features.

5.2 The solubility of oxygen

Two dissolved gases, oxygen and carbon dioxide, play major roles in the maintenance of life in aquatic systems. Oxygen is used by both animals and plants, whilst carbon dioxide is required by photosynthetic algae. If dissolved oxygen were not present in water many life forms would not survive. Dissolved oxygen levels vary with temperature, and oxygen is consumed in the degradation of organic matter in the water. Many reports of fish deaths in rivers turn out not to be the direct result of pollution but due to oxygen deprivation brought about by temperature fluctuations and/or the release of biodegradable organic matter, such as farm waste, into the water. In order to be able to assess the status of a water system it is necessary to understand something about the transport of gases in and out of a body of water. In the context of this chapter, which discusses the oxidizing and reducing properties of the aqueous environment, we will be particularly interested in the behaviour of oxygen.

There is no naturally occurring process which generates oxygen by purely chemical means, and all the oxygen which is present in water must therefore come from photosynthesis or from the atmosphere. Neither of these processes can be considered to be particularly effective at replacing lost oxygen.

This is another example of a situation in which the concentration of a species can be restricted by kinetic, rather than thermodynamic, considerations.

Photosynthetic organisms generate oxygen during the day but then consume it at night. The transfer of oxygen across an air–water interface is very slow in a static system but is greatly increased by turbulence (waves etc.).

We can calculate the equilibrium concentration of oxygen in water at 25 °C using the Henry's Law constant: 20.95 per cent of air (by volume) is oxygen and the partial pressure of oxygen in air, p_{O_2}, is therefore 0.2095 atmosphere. From Henry's Law:

$$[O_{2aq}] = K_H \cdot p_{O_2} = 1.28 \times 10^{-3} \times 0.2095$$

giving a dissolved oxygen concentration of 2.68×10^{-4} mol dm^{-3}. This converts to an equilibrium concentration of dissolved oxygen in water of 8.58 mg dm^{-3} at 25 °C.

The dissolved oxygen content of water is a major indicator of its oxidizing or reducing characteristics. We will now investigate how this is likely to influence the chemical forms of the elements found in natural and polluted waters.

The equilibrium concentration of dissolved oxygen in a water sample is highly temperature dependent dropping to 7.03 mg dm^{-3} at 35 °C. In hot weather fish deaths often result from oxygen starvation arising from the poor solubility of oxygen at the elevated temperature and the slow rate at which utilized dissolved oxygen can be replaced (kinetic control).

5.3 Oxidation and reduction

Oxidation is the removal of an electron from a chemical species. When the manganese(II) or manganous ion, for example, is oxidized to manganese(IV), two electrons are removed:

$$\text{Mn(II)} - 2e^- \rightleftharpoons \text{Mn(IV)}$$

At most natural pH values the manganese(II) will be present as Mn^{2+}, which is soluble. Mn(IV), however, will be precipitated as hydrated manganese(IV) oxides.

Reduction is the reverse of the oxidation process, resulting from the addition of electrons to the chemical species. Again this can result in radical changes in the properties of an element. The reduction of dissolved Hg^{2+} to metallic Hg^0, for example, produces a volatile product which can move from solution into the atmosphere.

Oxidation and reduction (redox) are controlled by the concentrations of electrons which are present. The concentration of electrons can be written as $[e^-]$ and given in terms of pE:

$$\boxed{pE = -\log_{10}[e^-]}$$

The electron only differs from other chemical species in solution by being rather small. To consider it as being different could be considered to be sizeist. Its apparent concentration in solution can be quoted as readily as that of any other species.

High pE implies a low concentration of electrons and the greatest tendency to oxidize. Low pE is equivalent to a high concentration of electrons and is therefore reducing.

The concentration of electrons in solution is most readily measured as an electrode potential which develops when the sample is incorporated into an electrochemical cell. The most common measure of redox activity is E_H, the electrode potential measured against the hydrogen electrode (Fig. 5.1). The unit of E_H is the volt.

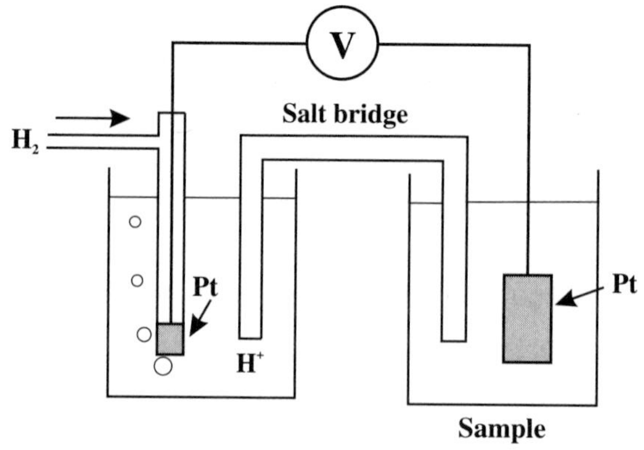

Indirect chemical measures of redox potential are often employed in field studies to overcome practical problems involved in the use of the apparatus shown in Fig. 5.1.

Fig. 5.1 The measurement of redox potential against the hydrogen electrode.

The E_H is related to the standard redox potential E_H^0 by the Nernst equation:

The Π in this equation indicates the product of the terms which follow (cf. the use of Σ to imply the sum). In this equation each concentration must be raised to powers (m and n) derived from the reaction equation.

It is often useful to know that $2.303RT/F = 0.059$ at 25 °C.

$$E_H = E_H^0 + \frac{RT}{nF}\log_e \frac{\Pi[\text{Oxidizing species}]^m}{\Pi[\text{Reducing species}]^n}$$

where

R = universal gas constant
T = temperature (degrees Kelvin)
n = the number of electrons transfered
F = Faraday's constant

The redox potential of a water sample is largely governed by the dissolution of oxygen and the utilization of the gas by organisms and chemicals (natural and man-made). Whether a body of water is oxidizing or reducing in character in turn determines the chemical forms of the trace elements which are in it. For some elements a change in oxidation state can result in almost complete precipitation or dissolution of the element and its transfer from the water to the sediments or vice versa. As mentioned previously, the oxidation of manganese(II) to manganese(IV), for example, results in the formation of the highly insoluble hydrated manganese oxides which precipitate out of solution and fall to the sediments. Oxidation of Fe(II) to Fe(III) results in a similar effect with the precipitation of dissolved iron from solution as the bright orange–brown hydrated iron(III) oxide solid (normally written as $Fe(OH)_3$ or FeOOH).

The equation for the manganese oxidation is:

The subscript s signifies that the species is a solid.

$$Mn^{2+} + 2H_2O \rightleftharpoons \gamma\text{-}MnO_{2s} + 4H^+ + 2e^- \qquad K = 10^{-40.8}$$

treating the electron just like any other species gives us the equilibrium constant:

$$K = \frac{[H^+]^4[e^-]^2}{[Mn^{2+}]} = 10^{-40.8}$$

taking logs

Note that the solvent and solid phase species are both always omitted from the equation for K.

$$\log_{10}K = 2\log_{10}[e^-] + \log_{10}\left(\frac{[H^+]^4}{[Mn^{2+}]}\right)$$

$$-\log_{10}[e^-] = pE = 20.4 + \frac{1}{2}\log_{10}\left(\frac{[H^+]^4}{[Mn^{2+}]}\right)$$

This is just one example of the more general equation:

$$pE = pE^\circ + \frac{1}{n}\log_{10}\frac{\Pi[\text{Oxidizing species}]^m}{\Pi[\text{Reducing species}]^n}$$

If this is compared to the Nernst equation, the similarities will be evident:

$$E_H = E_H^\circ + \frac{RT}{nF}\log_e\frac{\Pi[\text{Oxid.species}]^m}{\Pi[\text{Red.species}]^n}$$

where

$$pE^\circ = \frac{1}{n}\log_{10}K$$

= relative electron concentration when all other species are at unit concentration

5.4 The natural limits of redox in natural waters

The oxidation and reduction of water (to oxygen or hydrogen respectively) limits the range of redox potentials which are possible in a natural water body. With a knowledge of the redox equilibria describing the reactions we can derive equations which define the most oxidizing and reducing conditions which are possible in water. When plotted on a stability field diagram having pE and pH axes (a pE–pH diagram), they provide boundary conditions outside which it is unlikely that a water body (at equilibrium) will exist.

The oxidation of water

The equation for the oxidation of water:

NB this equation contains e^- and H^+ terms and is therefore both pE and pH dependent.

$$2H_2O \rightleftharpoons O_2 + 4H^+ + 4e^- \qquad \log K = -83.1$$

gives us the equilibrium expression:

$$K = p_{O_2}[H^+]^4[e^-]^4$$

p_{O_2} is the partial pressure of oxygen.

having the log form:

$$pE = -\frac{1}{4}\log K + \frac{1}{4}\log p_{O_2} - pH$$

As the partial pressure of oxygen in water can not exceed 1, the most oxidizing conditions possible in water are therefore:

$$pE = \frac{83.1}{4} + \frac{1}{4}\log 1 - pH$$

$$\boxed{pE = 20.75 - pH}$$

This can be drawn on a pE/pH diagram as a boundary line:

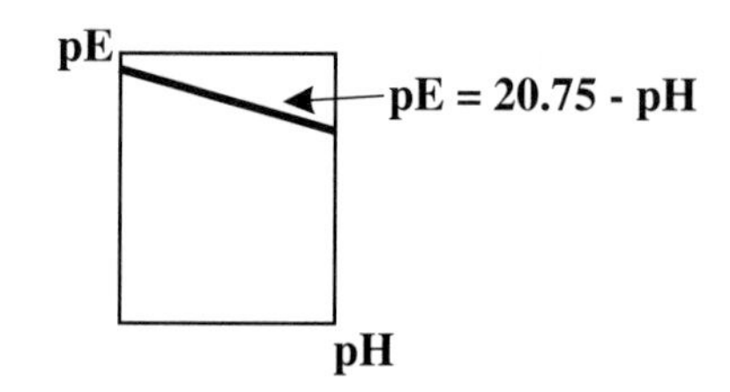

Under conditions above the line on the diagram i.e. when the pE is greater than 20.75 – pH, water itself will be oxidized.

The reduction of water

At the other end of the redox scale, the situation is reached at which the water is reduced:

NB the equation contains both e^- and H^+ terms and is therefore also pE and pH dependent.

$$2H^+ + 2e^- \rightleftharpoons H_2 \qquad \log K = 0$$

$$K = \frac{p_{H_2}}{[H^+]^2[e^-]^2}$$

$$pK = -\log p_{H_2} - 2pH - 2pE$$

$0 = -\log_{10}1 - 2pH - 2pE$

The maximum possible p_{H_2} is 1 (as before) and p*K*= 0, therefore

$$pE = -pH$$

When drawn on a pE–pH diagram, this line gives a boundary condition, below which the water will be reduced.

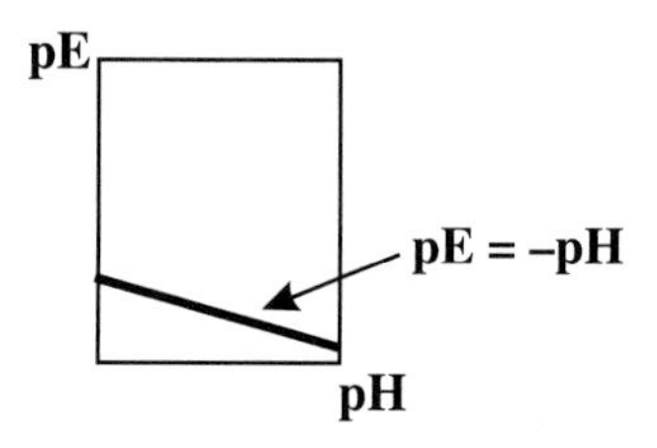

At pH 7 the maximum p*E* is: pE > 20.75 – 7.0 > 13.75 and the minimum: pE < –7.0

Together the oxidation and reduction equations give the limits outside which it should not be possible to find water. At pH 7.0 therefore, the pE must lie between 13.75 and –7.0. Above a pE of 13.75 the water is oxidized and below –7.0 the water is reduced.

Two measures of redox activity, pE and E_H, have been encountered so far and these are linearly related by the equation:

Both pE–pH and E_H – pH diagrams will be found in the literature, they only differ in the scale used on the redox axis.

$$pE = \frac{F}{2.3RT}E_H$$

This allows us to rewrite the equations defining the limits of redox activity in water in terms of E_H and to draw a corresponding E_H–pH diagram.

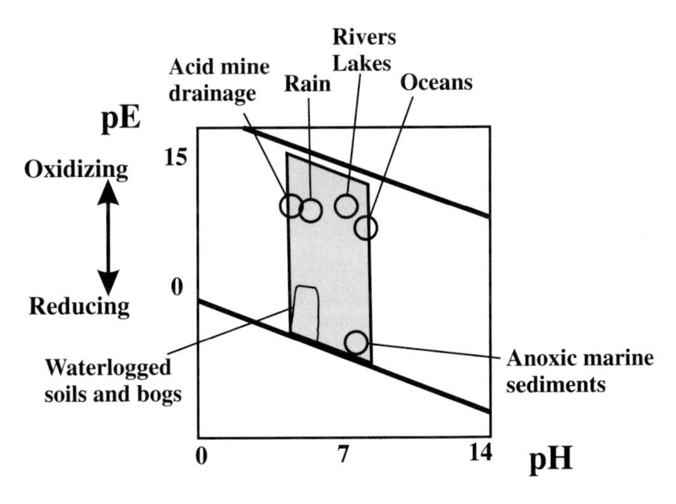

Fig. 5.2 Redox and acidity conditions of natural water types.

5.5 The redox conditions in natural waters

We are now in a position to be able to relate the chemistry occurring in a body of water to its origins and chemical condition.

Regions of the pE–pH diagram can be associated with the acidity and redox characteristics of particular types of water (Fig. 5.2) and from this association deductions can be made regarding the nature of the chemical species which are likely to be present in solution. The largest single water body, the World's oceans, has a relatively constant composition, is generally well oxygenated (high pE) and has a fairly consistent pH of *c.* 8.2. We can therefore represent this, and other water types, as areas on the pE–pH diagram.

Whilst well-oxygenated waters have high redox potentials (oxidizing), water that is depleted of oxygen will have more reducing characteristics. Such depletion occurs when biological or chemical processes use up the dissolved oxygen faster than it can be replenished. This will be the situation in the sub-surface water in peat bogs, some ground waters, and in the interstitial waters of anoxic sediments. A wide range of acidities are also to be found, ranging from pH values below 2 in volcanic areas to the highly alkaline conditions of some lakes.

5.6 pE–pH stability field diagrams

In the previous chapter the stability field diagram was introduced as a particularly useful way of illustrating the species which are present in systems which depend on two variables. Fig. 5.3, for example, shows the major mercury species in a simple aqueous system under varying conditions of acidity and redox potential.

The diagram is like an aerial map of a farm showing the crops which are growing in the fields; the lines on the diagram are the fences between the fields. As we move from point A to point B in the diagram we go from a field

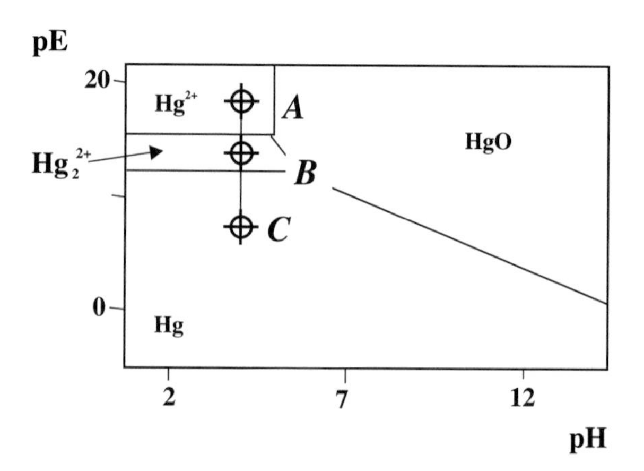

Fig. 5.3 A simple stability field diagram showing the predominant mercury species as functions of acidity and redox potential.

Notice how on moving from point A to C the oxidation state is reducing from two at point A, through one at B to metallic mercury (oxidation state zero) at C.

in which the dominant species in solution is Hg^{2+} to one in which ${Hg_2}^{2+}$ predominates. On the way, the concentration of Hg^{2+} gradually drops whilst the concentration of ${Hg_2}^{2+}$ increases. At the boundary line the concentrations of these two species are equal – the position of this can be found from the equation which relates the concentrations of the species in the two adjoining fields.

The best way to understand such a diagram is to partially derive it, the example to be worked through here involves the redox chemistry of iron, a common and often crucial component of many environmental systems. The chosen example will demonstrate how both acidity and redox combine to influence the species which are present in a natural water and how such changes can result in the precipitation and dissolution of iron.

To begin the construction of the diagram we will start with the conversion of Fe^{2+} to Fe^{3+}.

The oxidation of Fe^{2+} to Fe^{3+}.

The equation which relates the concentration of ferrous (Fe^{2+}) to ferric (Fe^{3+}) iron is:

$$Fe^{3+} + e^- \rightleftharpoons Fe^{2+} \quad \log K = +13.2$$

$$K = \frac{[Fe^{2+}]}{[Fe^{3+}][e^-]}$$

Taking logs

$$\log K = \log\frac{[Fe^{2+}]}{[Fe^{3+}]} - \log [e^-]$$

$$-pK = -\log\frac{[Fe^{3+}]}{[Fe^{2+}]} + pE$$

$$pE = 13.2 + \log\left(\frac{[Fe^{3+}]}{[Fe^{2+}]}\right)$$

NB no H^+ in the equation, therefore not pH dependent

When the concentration of Fe^{2+} equals that of Fe^{3+} we have the boundary between the Fe^{2+} and the Fe^{3+} regions. This occurs when $[Fe^{3+}] = [Fe^{2+}]$ making $\log ([Fe^{3+}]/[Fe^{2+}]) = 0$ and giving:

$$\boxed{pE = 13.2}$$

This equation tells us that at pE values below 13.2, the major form of iron is Fe^{2+}, and above pH 13.2 the dominant species becomes Fe^{3+}. We can draw this line on our stability field diagram but cannot, at present, be certain where it starts or ends.

NB this line is parallel to the pH axis as there is no pH dependency for this reaction.

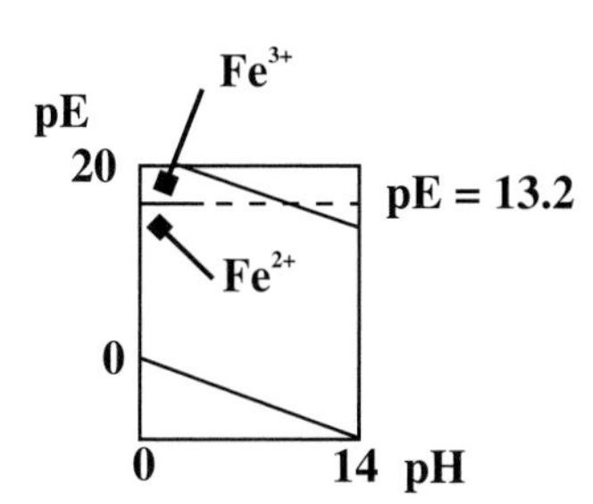

The oxidation to Fe^{3+} results in a species which readily undergoes hydrolysis to give a hydrous iron oxide precipitate ('$Fe(OH)_3$').

For a hydrolysis to happen OH^- must be present in sufficient quantity. If Fe^{3+} is to exist without hydrolysis occurring, the pH will have to be low (i.e. $[OH^-]$ is low).

Hydrolysis of Fe^{3+}

At higher pH the hydrolysis results in the formation of the insoluble iron(III) hydroxide (hydrous iron(III) oxide). How high does the pH have to be for this to occur?

The boundary between Fe^{3+} and $Fe(OH)_{3s}$ is given by

$$Fe(OH)_{3s} + 3H^+ \rightleftharpoons Fe^{3+} + 3H_2O$$

Which has an equilibrium constant

$$K_S = \frac{[Fe^{3+}]}{[H^+]^3} = 9.1 \times 10^3$$

NB $Fe(OH)_3$ is a solid phase species and does not therefore feature in the K_S equation.

which on taking logs and rearranging gives:

$$pH = 1.32 - \frac{1}{3}\log[Fe^{3+}]$$

$$\log K_S = \log [Fe^{3+}] - 3\log[H^+]$$

$$pH = \frac{1}{3}\log K_S - \frac{1}{3}\log[Fe^{3+}]$$

Assuming a maximum soluble Fe^{3+} concentration of 10^{-5} mol dm^{-3}

$$\boxed{pH = 2.99}$$

Transposed onto our pE–pH diagram this equation defines the boundary between the regions dominated by Fe^{3+} and $Fe(OH)_3$.

NB this boundary is not determined by the redox characteristics of the system and is therefore independent of pE. It therefore runs down the graph parallel to the pE axis.

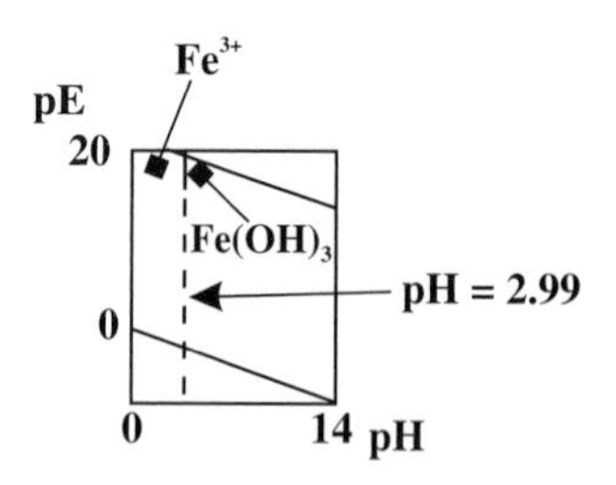

This precipitation in a natural system results in deposition of the iron into the sediments.

Hydrolysis of Fe^{2+}.
This is almost identical to the Fe^{3+} hydrolysis. The Fe^{2+}/$Fe(OH)_2(s)$ boundary is given by:

$$Fe(OH)_2 + 2H^+ \rightleftharpoons Fe^{2+} + 2H_2O$$

from the expression for the solubility

$$K_S = \frac{[Fe^{2+}]}{[H^+]^2}$$
$$= 8 \times 10^{12}$$

and assuming the concentration of Fe^{2+} to be 10^{-5} mol dm^{-3}

$$\log K_S = -2\log[H^+] + \log[Fe^{2+}]$$
$$pH = \frac{1}{2}\log K_S - \frac{1}{2}\log[Fe^{2+}]$$
$$= 6.45 + \frac{1}{2} \cdot 5$$

$$\boxed{pH = 8.95}$$

Which on the pE–pH diagram is a vertical line:

Again independent of pE and therefore parallel to the pE axis.

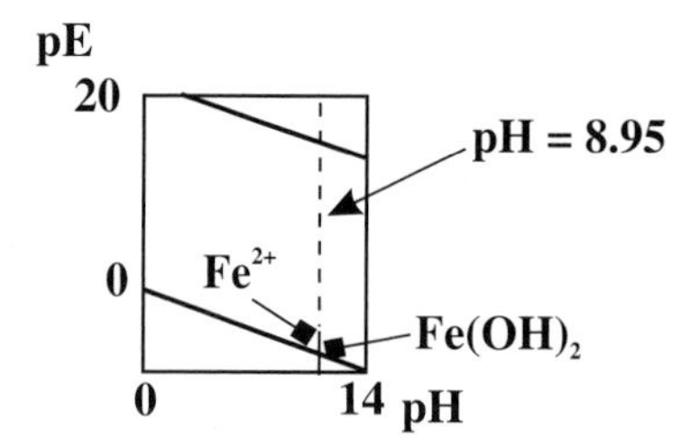

Interconversion of $Fe(OH)_3$ and Fe^{2+}.
The final line we are going to derive is the boundary between $Fe(OH)_3$ and Fe^{2+}. This can be considered to be made up from an oxidation of Fe^{2+} to Fe^{3+} which is then hydrolysed. Combining equations for these two steps we can get the equation for the overall reaction describing the conversion of Fe^{2+} to $Fe(OH)_3$.

$$Fe^{2+} \rightleftharpoons Fe^{3+} + e^- \quad \log K = -13.2$$
$$Fe^{3+} + 3H_2O \rightleftharpoons Fe(OH)_3 \quad \log K = -4.0$$

add the two equations for the conversion of Fe^{2+} into $Fe(OH)_3$

$$Fe^{2+} + 3H_2O \rightleftharpoons Fe(OH)_3 + 3H^+ + e^- \quad \log K = -17.2$$

$$K = \frac{[H^+]^3[e^-]}{[Fe^{2+}]}$$

and treating as before gives:

$$\log K = 3\log[H^+] + \log[e^-] - \log[Fe^{2+}] = -17.2$$
$$pE = 17.2 - \log[Fe^{2+}] - 3pH$$

$$pE = -\log[e^-]$$
$$= 3\log[H^+] - \log K - \log[Fe^{2+}]$$
$$= -3pH - \log K - \log[Fe^{2+}]$$

Assuming $[Fe^{2+}] = 10^{-5}$ mol dm^{-3} gives

$$\boxed{pE = 22.2 - 3pH}$$

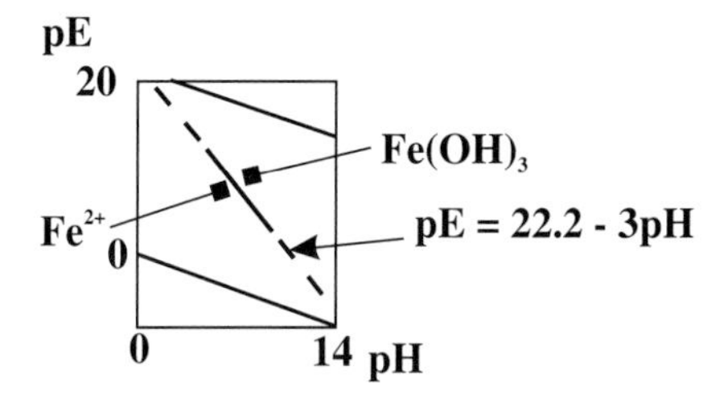

This combines elements of redox and pH dependency and is therefore a diagonal across the graph.

Once a few more equilibria have been incorporated, a full pE–pH diagram can be constructed (Fig. 5.4).

Carbonate will originate from dissolved carbon dioxide from the atmosphere and from the dissolution of carbonate rocks such as limestone.

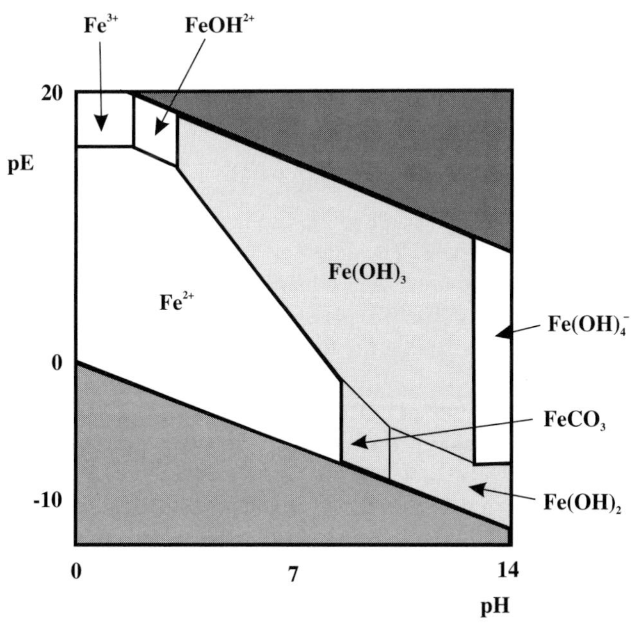

Fig. 5.4 A pE–pH diagram for iron incorporating carbonate species.

Once all the boundary lines have been established they define areas in which particular species predominate. Some of these species are readily soluble whilst others are normally insoluble species. Those species which are generally considered to be solid phase species are shaded in the diagram. Now that the diagram is complete it will be possible to identify a number of trends. These can be viewed from two directions.

From left to right of the diagram are the pH derived changes. The non-complexed, hydrated ions (Fe^{2+} and Fe^{3+}) are well over to the left. Such species can only exist under acidic conditions as elsewhere on the diagram the concentration of OH^- ions is too high. The right of the diagram, the high pH region, is the area in which $OH^-$is competing most strongly as a ligand. The number of OH^-groups per iron therefore increases left to right. Notice how the comparatively weak carbonate complex is only important within a narrow pH range in the centre of the graph. At higher pH the OH^- ligand successfully competes against the carbonate for the metal centre. At low pH the ligand is protonated and can not coordinate to the metal.

$$CO_3^{2-} \overset{H^+}{\rightleftharpoons} HCO_3^- \overset{H^+}{\rightleftharpoons} H_2CO_3$$

Redox behaviour changes from the top to the bottom of the diagram; oxidizing conditions at the top and reducing at the bottom. The highest oxidation state species, such as iron (III) are therefore to be found at the top of the diagram whilst low oxidation state species are at the bottom.

The diagram can be used to trace the behaviour of iron in natural waters. Under acid reducing conditions which are to be found in some groundwaters (bottom left), Fe^{2+} is the major species. Unless we go to very low pH and very high pE, oxidation of the iron(II) will result in the formation of $Fe(OH)_3$—a species which is essentially insoluble. Changes of redox conditions can therefore have radical effects on the chemical form and availability of an element. In this particular example it is governing whether the iron will be in solution or in the sediments!

5.7 Case studies

5.7.1 Iron transport through soil

As rainwater permeates through soil it can leach metals from the surface and carry them to deeper levels. The extent of this transfer of metal from the soil surface depends on the prevailing conditions and the chemical properties of the individual elements. The pE–pH stability field diagram developed in this chapter can be used to explain the unusual distribution of iron in some soils.

Let us assume that the pH of rain is 4.5 and that it has a pE of 10 as it starts to permeate through the surface of an organic-rich soil (Fig. 5.5, Point A).

$$2H^+ + CO_3^{2-} \rightleftharpoons H_2CO_3 \rightarrow CO_2$$

As the water moves deeper, the organic content of the soil is reduced and the acidity drops as it is neutralized by soil components such as carbonates. In the absence of redox changes, if the pH were to rise to 6 this would take us to Point B on the diagram where the predominant species is $Fe(OH)_3$. This is a highly insoluble species which would be unable to migrate further.

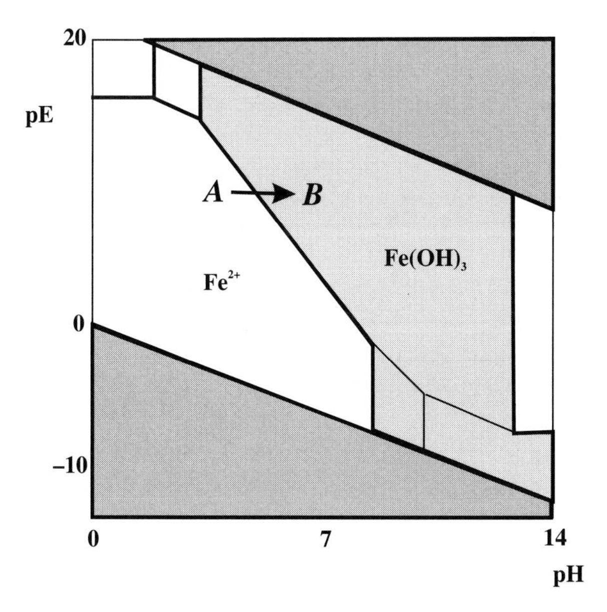

Fig. 5.5 Partial iron pE–pH diagram.

This is the reason for the formation of the so-called 'podzol' soils, typical of the forest soils of humid parts of the northern temperate zones. Vegetation is abundant in these areas giving a soil water pH of between 3.5 and 4.5. The acid water leaches out the alkali and alkaline earth ions from the surface soil together with much of the iron and aluminium. The layer left behind is rich in silica and a grey–white colour.

'Podzol' is Russian for 'under ash', a description of the appearance of this layer.

The iron and aluminium, once in solution, are mobile and can be carried downwards (Fig. 5.6). As the pH of the water rises however, hydrous iron oxide and colloidal clay minerals are deposited, resulting in an iron and clay rich horizon called a 'clay hardpan'.

The clay is typically kaolinite—expected from the acid environment and depleted of calcium and magnesium.

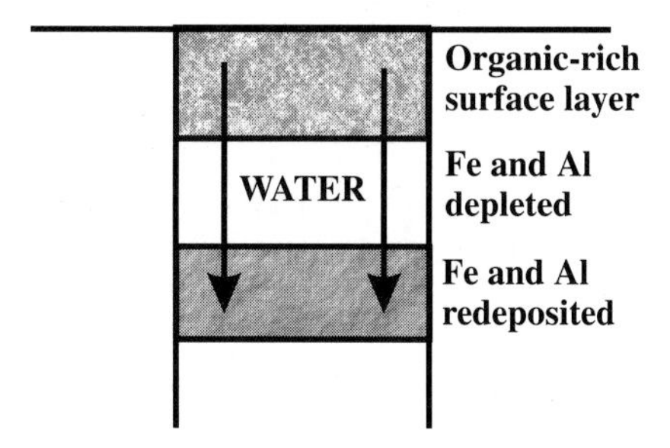

Fig. 5.6 Metal leaching and redeposition in soil.

In some circumstances, such as in a waterlogged soil, the transport of oxygen to the lower levels becomes difficult. The little oxygen which is present is used for microbial respiration and the pE might as a result drop to 0

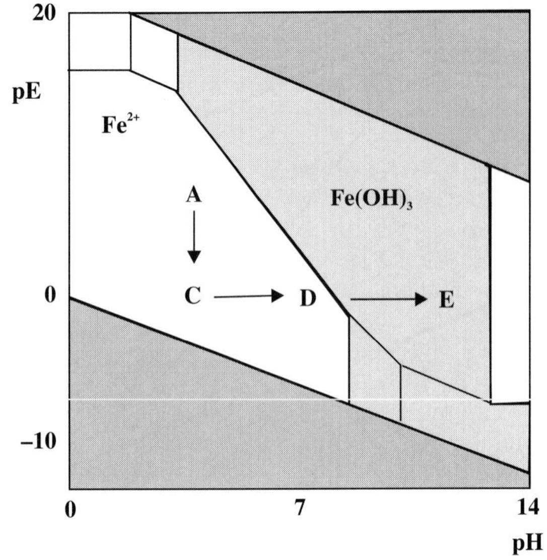

Fig. 5.7 pE–pH diagram for iron species in a waterlogged soil.

(Point C, Fig. 5.7). As the water permeates deeper into the soil the pE remains fixed but, as before, the pH might rise.

Whereas in the well oxygenated situation the $Fe^{2+}/Fe(OH)_3$ boundary was crossed at a pH of approximately 5, under the anoxic conditions a pH of over 8 (Point D) is required for the deposition of the $Fe(OH)_3$. This results in the migration of the iron over longer distances under reducing conditions.

5.7.2 The acidification of coal mine drainage waters

The main sulfur-bearing mineral which is found in coal seams is the iron sulfide ore pyrite (Fig. 5.8). For simplicity we will represent this as FeS_2.

Despite initial appearances, FeS_2 has iron in the +2 oxidation state, it is a salt of the disulfide ion S_2^{2-}.

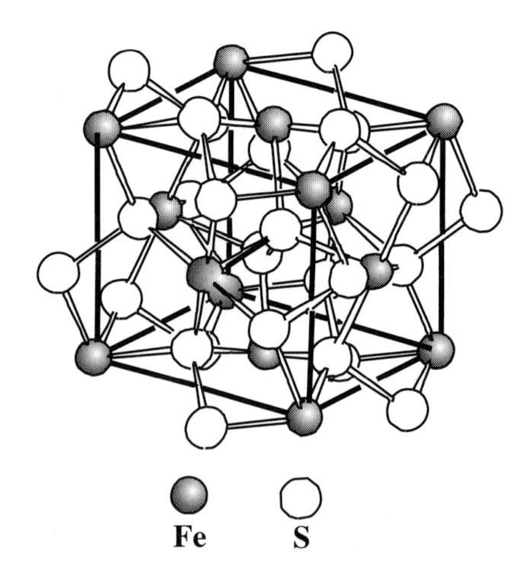

Fig. 5.8 The structure of pyrite.

During coal mining operations, pyrite is exposed to air and water and the following overall reaction occurs:

Notice how (i) the sulfide (S(-II)) has oxidized to sulfate (S(VI)) and (ii) the iron has oxidized to Fe(III).

$$4FeS_2 + 15O_2 + 14H_2O \rightarrow 4Fe(OH)_3 + 16H^+ + 8SO_4^{2-}$$

This reaction results in the production of four protons for every pyrite molecule which is oxidized. The result is a significant lowering of the pH of water draining a mine. A number of reactions are believed to be taking place which lead to this acidity.

In some mines leather boots have been known to rot in less than a week.

The first reaction is the oxidation of the pyrite sulfur, releasing the Fe^{2+} into solution:

Iron(II) salts are generally soluble and remain in solution.

$$FeS_2(s) + \frac{7}{2}O_2 + H_2O \rightarrow Fe^{2+} + 2SO_4^{2-} + 2H^+ \quad (7.8)$$

The rate limiting step, with a half time of *c.* 1000 days at pH 3, is the oxidation of Fe^{2+} to Fe^{3+}:

Iron(III) salts generally hydrolyse to insoluble products and are deposited in the sediments.

$$Fe^{2+} + \frac{1}{4}O_2 + H^+ \rightarrow Fe^{3+} + \frac{1}{2}H_2O \quad (7.9)$$

The Fe^{3+} is then either hydrolysed

$$Fe^{3+} + 3H_2O \rightleftharpoons Fe(OH)_3(s) + 3H^+ \quad (7.10)$$

or becomes a secondary oxidant of sulfide to release more Fe^{2+} into solution:

$$FeS_2(s) + 14Fe^{3+} + 8H_2O \rightarrow 15Fe^{2+} + 2SO_4^{2-} + 16H^+ \quad (7.11)$$

Overall, for every mole of iron pyrite dissolved, four equivalents of acidity are released.

These mechanisms result in mine drainage waters becoming highly acidic and it is not unusual to find streams in the vicinity of a mine with pH values of 1 to 2. This is sufficiently acidic to result in the dissolution of other, often more toxic, metals from the mineral deposits. Streams in mining areas are often therefore 'naturally' contaminated with dissolved heavy metals such as lead, cadmium, mercury, and arsenic. As this acidic stream water moves away from the mine, the acidity gradually reduces due to reactions with soil and minerals, mixing with uncontaminated water, and in some cases discharge into an estuary. This increased pH results in the deposition from solution of those metals which are only soluble at very low pH.

The chemical explanation of acid mine drainage is not however the complete story as much of the acidity is believed to be due to biological oxidation of the sulfide ores. Bacteria of the *Thiobacillus* family use sulfur and atmospheric oxygen as an energy source resulting in the oxidation of sulfide to sulfate. This can be thought of as an exchange of the strongly basic anion S^{2-} for the neutral anion SO_4^{2-}, a conversion which results in an increase in the acidity of the water.

5.8 Problems

1. In calculating the oxidative redox limit for water in Section 5.4 it was assumed that the partial pressure of oxygen could not exceed 1. Calculate whether it would have made a significant difference to the result if a maximum partial pressure of 0.2 had been assumed?

2. The diagram below illustrates some oxidation/reduction chemistry of lead.

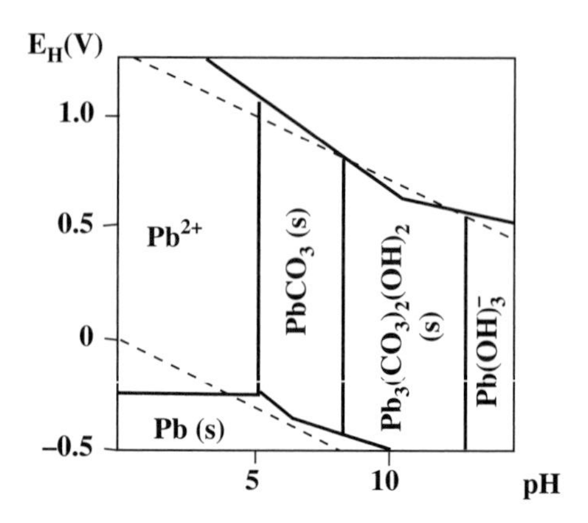

Using this diagram and any other pertinent information that you need to draw upon:

a) What are denoted by the two dashed lines on the diagram?

b) Estimate the pH and redox conditions in the following types of water:

   i) surface ocean water
   ii) anaerobic, acidic mine drainage waters
   iii) oxic pond sediments
   iv) a surface water of a lake which has been affected by acid rain.

For each of these water types identify the most stable lead species.

c) Water emerges as a spring from an anaerobic groundwater source with a pH of 4. It then flows over a limestone surface. Describe what would be observed, and explain the chemistry of what is happening.

3. The E_H-pH diagram shown below has been incorrectly labelled. Identify the errors and redraw the diagram correcting the mistakes where necessary. Explain the reasons behind your positioning of each new label.

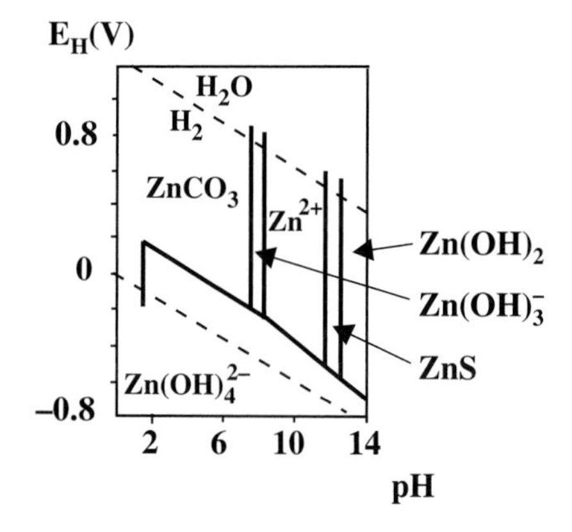

4. The diagram below shows the distribution of dissolved iron and manganese in a lake. The lake is approximately 37 metres deep and the top of the graph represents the water surface.

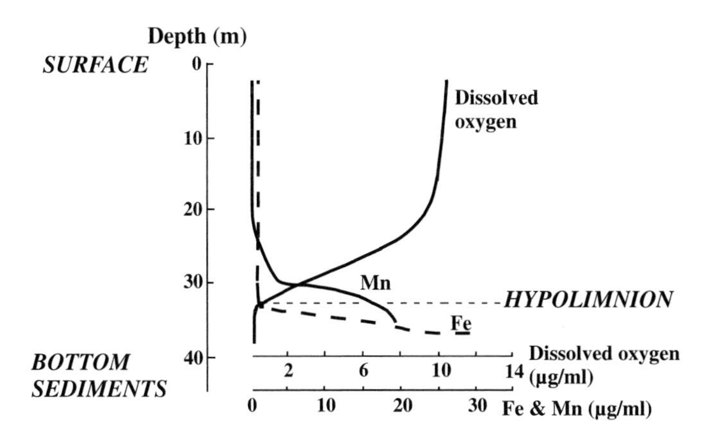

i) Discuss factors which influence the concentration of dissolved oxygen in the water column. Explain the distribution of dissolved oxygen in this lake.

ii) What are the major iron and manganese species which are likely to be present in the lake? How is the speciation of these elements indicated by the distribution of dissolved oxygen in the water column?

iii) What is the hypolimnion?

iv) Why are the concentrations of dissolved iron and manganese in the surface water very low?

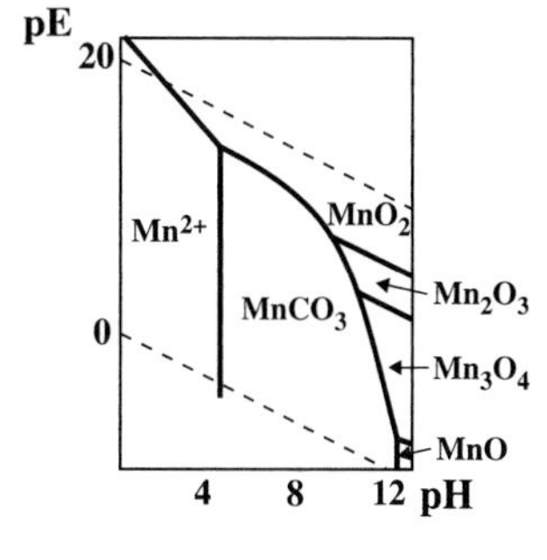

5. Whilst a large proportion of all saline water bodies are oxic throughout most of the water column, in some cases geological and meteorological factors come together to produce the right conditions for the water to become anoxic. Some fjords, for example, have a cross section which results in the trapping of a water mass, preventing it from freely mixing with the overlying oxygenated water (Fig. 5.12). This trapped water can readily become anoxic.

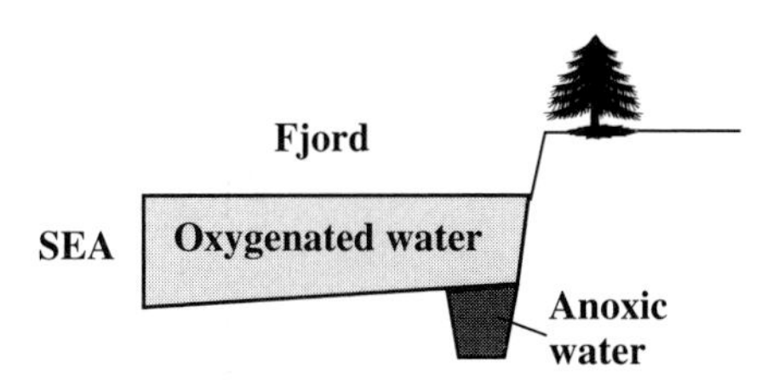

Fig. 5.12 A trapped anoxic water mass in a fjord.

In Saanich Inlet in Canada, a steep change in redox characteristics occurs at a depth of *c.* 100 m (Fig. 5.13). This results in an interesting distribution of chemical species in the water column.

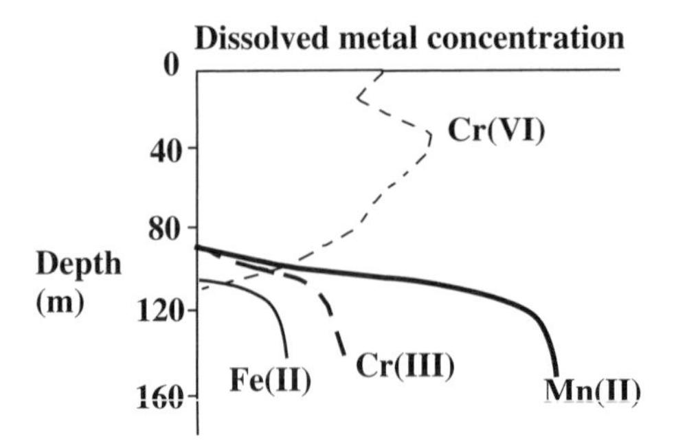

Fig. 5.13 The vertical distribution of dissolved metals in the Saanich Inlet water column.

HINT: the diagram shows the dissolved species. Where are the Fe(III) and Mn(IV) species?

Can you explain the distributions of iron, manganese, and chromium in the water column? The pE–pH diagrams for chromium and manganese are shown below; the iron diagram appears in Fig. 5.4.

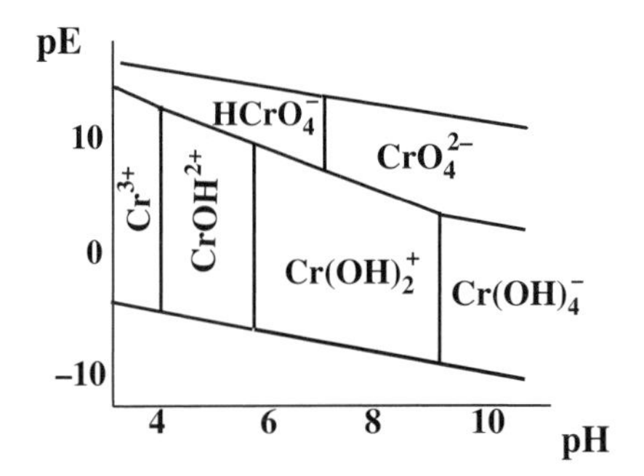

Fig. 5.14 pE–pH diagram for chromium.

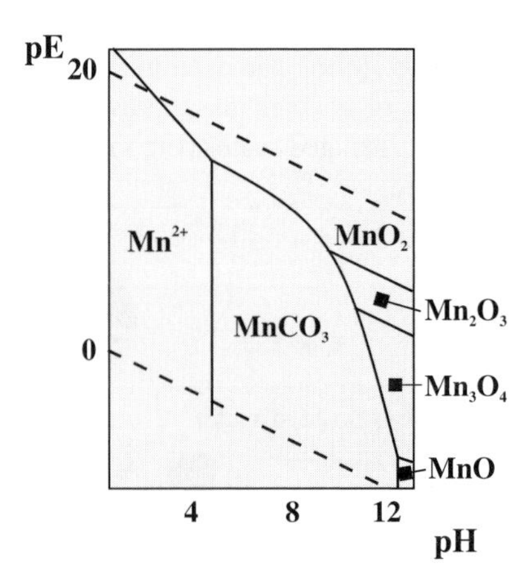

Fig. 5.15 pE–pH diagram for manganese species.

5.9 Further reading

Oxidation and reduction

Inorganic chemistry. Shriver, D. F., Atkins, P. W., and Langford, C. H. Oxford University Press, Oxford, 1994. ISBN 019855396X

Basic inorganic chemistry. Cotton, F. A., Wilkinson, G., and Gaus, P. L. Wiley, Chichester, 1995. ISBN 0471599743

Electrode potentials. Compton, R. G. and Sanders, H. W. Oxford Chemistry Primers. Oxford University Press, Oxford, 1996. ISBN 0-19-8556845

The diagrams

Environmental chemistry. Manahan, S. E. Lewis Publishers, Chelsea, Michigan, 1995. ISBN 0-412-484890-0

Aquatic chemistry. Stumm, W. and Morgan, J. J. Wiley-Interscience, New York, 1996. ISBN 0-471-51184-6

A classic advanced text.

Ionic equilibrium, a mathematical approach. Butler, J. N., Addison Wesley, Massachusetts, 1964.

Advanced and not easy but clear and well presented.

6 Dissolution and deposition processes

6.1 Introduction

No natural water body is a purely homogenous solution. As we have already seen, not only does water spend much of its time in contact with rocks, minerals, organisms, etc., but there is a constant interchange of material between the dissolved and solid phases. The composition of natural waters is rarely governed purely by homogeneous reactions, there is a continuous cycling of material between the dissolved and solid phases brought about by a combination of chemical, physical, and biological processes (Fig. 6.1).

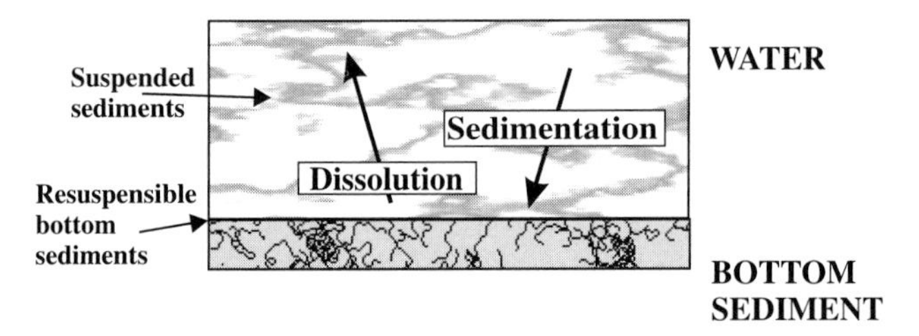

Fig. 6.1 The interchange of material between the sediments and water.

Some of the solid matter which is suspended in a natural water, or deposited to make up the top layers of the bottom sediments, is material which has been recently deposited from solution. This may arise from a purely chemical process such as precipitation or colloidal aggregation, or may result from the production of solid biomass, such as that which results from the photosynthetic activity of algae. Some of this solid material will be rapidly returned to the dissolved phase by changes in solution chemistry or bacterially assisted decomposition; the remainder will eventually be incorporated into the sediments.

Most sediments do not consist of one pure component but are made up of mixtures of clay, silt, sand, organic matter, minerals, and organic matter. They may be present as a result of local physical, chemical, or biological processes or may have been washed into the water body from external sources. The settlement of solid material out of solution onto the floor of a lake or ocean results in the build up of layers of sediment. If this is not disturbed by movement of the water, distinct layers can result giving a history of the sediment deposition. In other more turbulent regions, such as the mouths of estuaries, the energy of the incoming tide can result in the remobilization of significant depths of sediment during each tidal cycle.

6.2 Solubility and precipitation

Contact between the minerals, rocks and water results in a constant interchange of material between the dissolved and solid phases.

Sometimes this interchange is at equilibrium, other times not. The solution can therefore be supersaturated, undersaturated, or at equilibrium with the solid.

In the simplest of situations, when a salt dissolves resulting in the release of a cation and anion which do not subsequently react in solution, the extent of the dissolution can be described by a special equilibrium constant, the solubility product. For the dissolution of calcium fluoride:

$$CaF_2 \rightleftharpoons Ca^{2+} + 2F^-$$

like any other equilibrium, we can describe it using an equilibrium constant:

$$K = \frac{[Ca^{2+}][F^-]^2}{[CaF_2]}$$

The concentration of exposed calcium fluoride is essentially unaffected by this dissolution process and we can therefore consider that $[CaF_2]$ is constant. The equilibrium constant can therefore be written as the solubility product K_s.

$$K_S = [Ca^{2+}][F^-]^2$$

NB the terms are raised to powers determined by the compound stoichiometry.

Calcium sulfate

A commonly encountered example of the movement of material between the dissolved and solid phases is the dissolution of calcium sulfate from rocks such as gypsum or anhydrite. When water flows through the rocks calcium sulfate dissolves slightly, giving free ions:

$$CaSO_4 \rightleftharpoons Ca^{2+} + SO_4^{2-}$$

the solution eventually becomes saturated when an equilibrium has been established. For $CaSO_4$, at 25 °C, K_s has a value of 3.4×10^{-5}.

At this point $CaSO_4$ is being deposited from solution as fast as it is dissolving.

If we dissolve $CaSO_4$ in pure water, one calcium ion is produced for every sulfate ion; their concentrations will therefore be the same and we can therefore write:

$$[Ca^{2+}] = [SO_4^{2-}]$$

making the solubility equation

$$K_S = [Ca^{2+}]^2 = 3.4 \times 10^{-5}$$

Giving a Ca^{2+} concentration of 5.8×10^{-3} mol dm^{-3}

This is rather too simplistic for most natural waters as these normally also contain calcium and sulfate ions from other sources. The solubility product

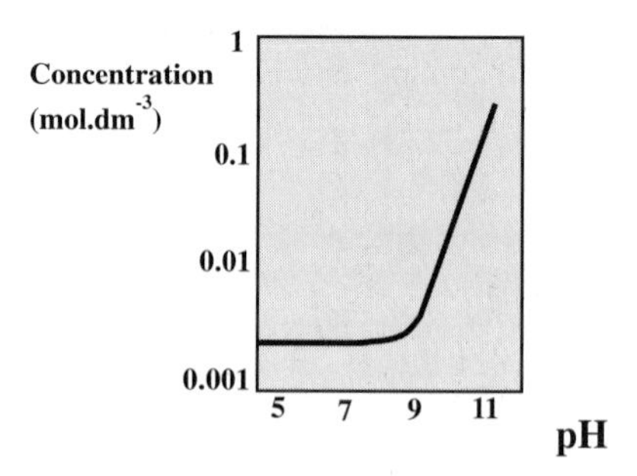

Fig. 6.2 The solubility of silica as a function of pH.

Put another way, the solubility of the calcium sulfate falls as the amount of sulfate ion increases.

equation still holds true, but if we were to add some sulfate ions to a $CaSO_4$ slurry in water, the concentration of calcium ions in solution would have to decrease to maintain K_s constant. This is often called **the common ion effect**.

Silica—solubility and pH

Amorphous silica provides another simple example of a dissolution process. The solubility of silica is pH dependent (Fig. 6.2).

The low pH part of Fig. 6.2 shows the concentration of dissolved silica remaining essentially constant with pH. In this range the silica which is in solution is mainly in the H_4SiO_4 form, resulting from the dissolution:

There is no H^+ term in this equation and the dissolution is therefore independent of pH.

$$SiO_2 + 2H_2O \rightleftharpoons H_4SiO_4$$

For which

$$K_S = [H_4SiO_4] = 1 \times 10^{-2.7} \text{ at } 25\ °C$$

NB the concentration of SiO_2 in solution does not depend on the quantity of silica in contact with the solution.

A solution containing an H_4SiO_4 concentration of greater than *c.* $10^{-2.7}$ mol dm^{-3} would therefore be supersaturated and on thermodynamic criteria the silica would be expected to precipitate out of solution. This could however under some circumstances be a slow process, i.e. kinetically controlled.

Charged species are normally more soluble than uncharged ones and deprotonation of the H_4SiO_4 would therefore be expected to increase its solubility.

At higher pH values the H_4SiO_4 deprotonates:

$$H_4SiO_4 \rightleftharpoons H_3SiO_4^- + H^+$$

for which

This equation contains an H^+ term. The $H_3SiO_4^-$ concentration will therefore depend on pH.

$$K_1 = \frac{[H_3SiO_4^-][H^+]}{[H_4SiO_4]} = 10^{-9.9}$$

At pH values above about 9.9 most of the H_4SiO_4 is therefore deprotonated and other species, such as $H_3SiO_4^-$, become important causing the solubility to increase with pH.

The deprotonation does not stop with the loss of one proton and at pH values above 11.7 the loss of a second proton becomes significant:

The acid dissociation constant pK_2 is 11.7.

$$K_2 = \frac{[H_2SiO_4^{2-}][H^+]}{[H_3SiO_4^-]} = 10^{-11.7}$$

Below pH 11.7 $H_3SiO_4^-$ will therefore be the predominant species. Above pH 11.7 there will be more $H_2SiO_4^{2-}$ than $H_3SiO_4^-$.

The solubility of ferrous (iron(II)) iron

The solubilities of many chemical species are governed by solution pH and a diagrammatic representation of such behaviour can often assist us in our understanding of the behaviour of material in the aquatic environment. Take, for example the case of iron(II) solubility. The solubility curve is shown in Fig. 6.3.

The shaded area of the diagram represents the conditions under which the iron(II) is precipitated as $Fe(OH)_2$. Outside that region the iron(II) is soluble. Within the unshaded part of the diagram the iron(II) is present as a variety of different species. At low pH the solution species is predominantly Fe^{2+}, at high pH $Fe(OH)_3^-$. The solubility of the iron is at its highest when in solution as Fe^{2+}, dropping down to a minimum at approximately pH12 before rising again at higher pH as the iron goes back into solution as $Fe(OH)_3^-$.

This is an example of amphoteric behaviour, the iron being soluble in both acidic and alkaline solution.

The construction of this graph from the equilibrium data therefore involves calculating the concentration of each of the soluble species (Fe^{2+}, $FeOH^+$, and $Fe(OH)_3^-$) when it is in equilibrium with solid phase $Fe(OH)_2$. In each case the equilibrium equation required is that which relates the species of interest to $Fe(OH)_2$.

With Fe^{2+} the appropriate equation is:

$$Fe(OH)_2(s) \rightleftharpoons Fe^{2+} + 2OH^- \qquad \log K_S = -14.7$$

$$K_S = [Fe^{2+}][OH^-]^2$$

but $[OH^-] = K_W/[H^+]$, gives

$$[Fe^{2+}] = \frac{10^{-14.7}}{[OH^-]^2} = \frac{10^{-14.7}[H^+]^2}{(10^{-14})^2} = 10^{13.3}[H^+]^2$$

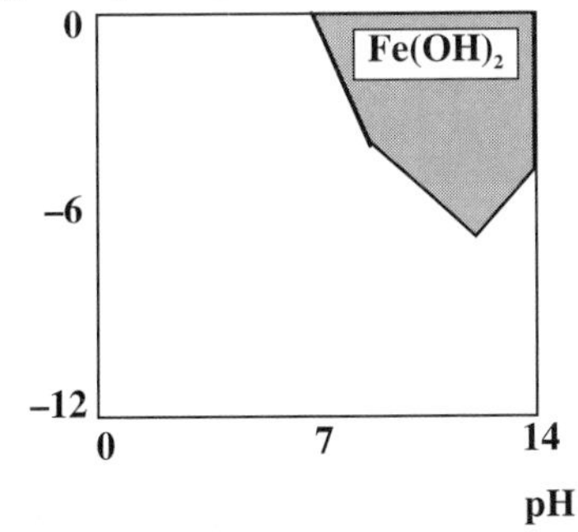

Fig. 6.3 Solubility of iron(II) species.

Taking logs:

$$\log[Fe^{2+}] = 13.3 - 2\text{pH} \qquad \text{(A)}$$

How the concentration of the Fe^{2+} changes with pH can be illustrated graphically:

This is a boundary line. To the left of line A the iron is soluble, to the right insoluble.

The formation of the dissolved species $Fe(OH)_3^-$ can be accounted for by the equation:

$$Fe(OH)_2(s) + OH^- \rightleftharpoons Fe(OH)_3^- \qquad \log K_S = -5.1$$

$$K_S = \frac{[Fe(OH)_3^-]}{[OH^-]}$$

treating as before:

$$\log[Fe(OH)_3^-] = -19.1 + \text{pH} \qquad \text{(B)}$$

Plotting this gives us line B

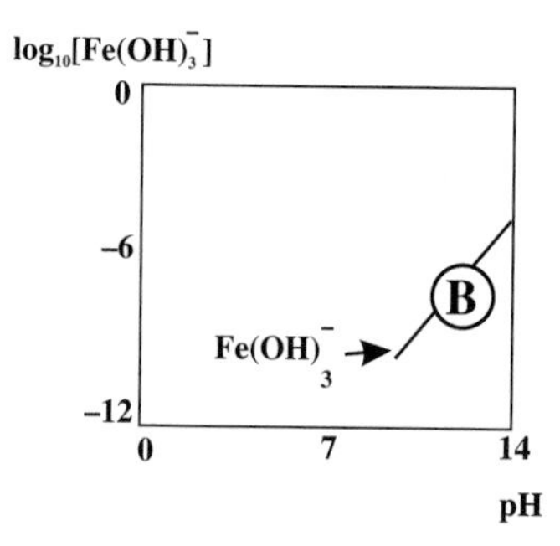

The loss of one OH^- from $Fe(OH)_2$ gives $FeOH^+$:

$$FeOH_2 \rightleftharpoons FeOH^+ + OH^- \quad \log K_S = -9$$

$$K_S = [FeOH^+][OH^-]$$

$$\frac{[FeOH^+]K_W}{[H^+]} = 10^{-9}$$

giving:

$$\log[Fe(OH)^+] = 5 - pH \qquad (C)$$

This is line C on the graph:

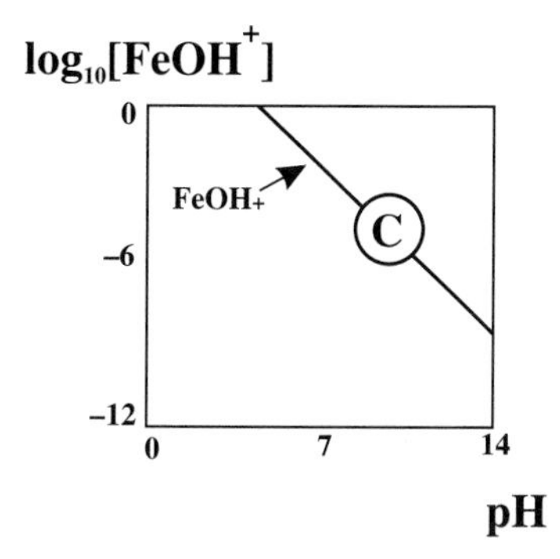

Each of these lines represents the limit of solubility of iron in one particular form. In the case of line C, the maximum concentration of iron that can be present in solution as $FeOH^+$. Putting all three lines on one diagram gives the solubility curve showing how the concentration of dissolved iron changes with acidity (Fig. 6.4).

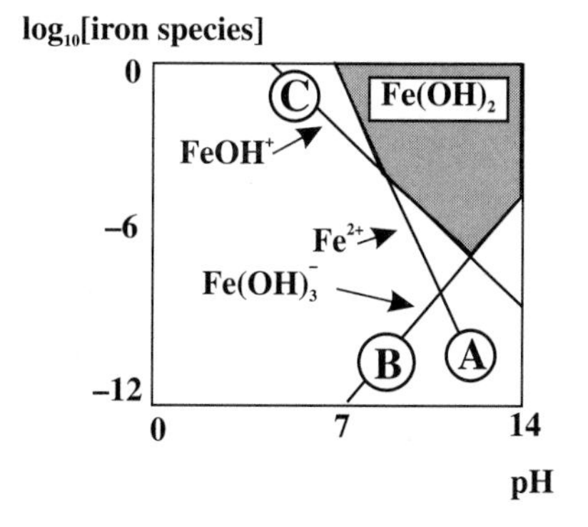

Compare with Fig. 6.3.

Fig. 6.4 Solubility of iron(II) species.

6.3 Chemical weathering

The intimate contact of rocks and minerals with water is a major factor in their erosion and the dispersal of solid material in soils and the river, lake, and coastal marine bottom and suspended sediments. The increased surface area resulting from the breakdown into smaller particles can enhance the rate of dissolution and change by exposing new surfaces to attack. This attack not only releases material into the water but can radically change the nature of the rocks and minerals. There are two main mechanisms which are responsible for these weathering effects, hydrolysis and oxidation.

Weathering by hydrolysis

Silicates weather primarily due to hydrolysis. In the case of the mineral forsterite this can be written as:

Most rivers, lakes, etc. are naturally slightly acidic due to the dissolved carbon dioxide from the atmosphere and respiration.

$$Mg_2SiO_4 + 4H_2O \Rightarrow 2Mg^{2+} + 4OH^- + H_4SiO_4$$

The hydrogen ions that are required for the reaction originate from the water, leaving the solution more alkaline after reaction.

Many silicates contain several cations and their hydrolysis is therefore more complex as the different cations go into solution at different rates. During the weathering the silicate grains are covered by an outer shell which has a different chemical composition to the bulk. This outer shell is formed by the preferential removal of some metals from the silicate and protects the core of the particle from further attack. The overall effect is to make the dissolution of most silicates extremely slow.

This weathering is therefore controlled by its kinetics. According to the thermodynamics the reaction should procede, it just takes a long while to get there.

The weathering of silicates does not necessarily result in complete dissolution. The alkaline earth metals may be removed leaving clay minerals. In the case of K-feldspar, kaolinite can be formed:

$$4KAlSi_3O_8 + 22H_2O \Rightarrow 4K^+ + 4OH^- + Al_4Si_4O_{10}(OH)_8 + 8H_4SiO_4$$

The resulting clay, kaolinite, has a very low solubility.

Weathering by oxidation

Iron minerals. The second most important mechanism of chemical weathering is oxidation. For this to occur at least one of the elements in the mineral structure must be in a lower oxidation state which can be readily oxidized. Whilst many elements fulfil this criterion the most commonly encountered examples contain iron(II), manganese(II), and sulfide.

Fe^{2+} oxidizes to Fe^{3+}; Mn^{2+} to Mn(IV) oxides; S^{2-} to SO_4^{2-}.

The weathering of iron minerals provides one of the most visible examples due to the noticeable orange–brown coloration of the iron(III) oxide product. Any ferrous (iron(II)) compound which is exposed to the air will be oxidized to its ferric (iron(III)) state. In the presence of water this results in the formation of a hydrated iron(III) oxide product which is often written as $Fe(OH)_3$ or $FeO(OH)$.

These formulae however imply a fixed stoichiometry, which is not the case and might be better represented as $Fe_2O_3.nH_2O$.

For the two iron minerals siderite and fayalite the overall result of the reaction can be represented as:

$$2FeCO_3(\text{siderite}) + \frac{1}{2}O_2 + 2H_2O \Rightarrow Fe_2O_3 + 2H_2CO_3$$

$$Fe_2SiO_4(\text{fayalite}) + \frac{1}{2}O_2 + 2H_2O \Rightarrow Fe_2O_3 + H_4SiO_4$$

In a natural water system this may involve sequential reactions:

$$Fe_2SiO_4 + 4H^+ \Rightarrow 2Fe^{2+} + H_4SiO_4$$

followed by the oxidation of the Fe^{2+}:

$$2Fe^{2+} + \frac{1}{2}O_2 + 2H_2O \Rightarrow Fe_2O_3 + 4H^+$$

Oxidation of sulfides. In igneous rocks, sulfur occurs in its −II oxidation state primarily as sulfides (S^{2-}). Exposure of these to air must eventually result in the oxidation of sulfide to sulfate (S(VI)):

$$ZnS + 2O_2 \Rightarrow Zn^{2+} + SO_4^{2-}$$

A more realistic mechanism might involve the formation of hydrogen sulfide under the acidic conditions provided by carbon dioxide dissolution, followed by the oxidation of the sulfide to sulfate.

In the case of lead sulfide, the lead sulfate product is fairly insoluble so we could write:

$$PbS + 2O_2 \Rightarrow PbSO_4$$

6.4 Colloids and their aggregation

The nature of colloids

Lying between the dissolved state, and particles which precipitate out of solution, there is a class of very small particles suspended in water, which range in diameter between 1 nm and 1 μm. These are colloids. They play a significant role in the aquatic chemistry of both inorganic and organic compounds. Because they are suspended in the water they are transported along with moving water bodies. Once destabilized however they rapidly come together to produce larger particles that can precipitate and drop to the bottom sediments, carrying with them other adsorbed and co-precipitated material.

The size of these particles is about the same as the wavelength of visible light and this results in white light scattering to give a light blue colour when viewed at right angles to the incident light (Tyndall effect).

The particles are not visible to the naked eye and a colloidal sol (the colloidal equivalent of a solution) appears completely homogeneous. When allowed to stand, colloids do not settle out like heavier particles.

Colloidal particles are charged and therefore move in an electric field. This primary charge may be positive or negative and may change with the solution pH. In aqueous solution however, there can be no overall charge and the primary charge must therefore be counterbalanced by the collection of opposite charges around the particles—a double layer therefore exists at the interface between the solid and the water (Fig. 6.5).

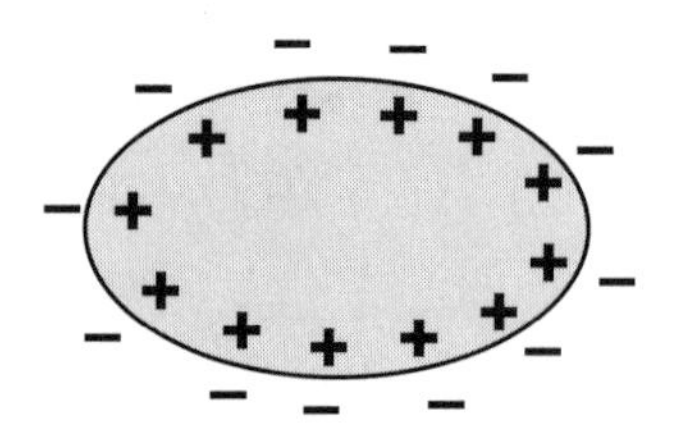

Fig. 6.5 A colloidal particle having a positive primary charge.

What keeps colloidal particles apart? Colloidal sols are sometimes particularly stable systems. Were it not for the repulsive forces between the particles it might be expected that the small particles of the colloid would come together (aggregate) and precipitate as larger particles. Under what conditions will a colloid remain stable and when we wish to remove colloidal material from waste water to purify it, how might the aggregation be induced?

When two particles approach, the electrostatic interaction between the particles of the same primary charge causes a repulsion which increases as the particles get closer (Fig. 6.6). At the same time van der Waal's forces are acting on the particles giving an attractive potential V_A which also increases as the particles come closer (Fig. 6.7).

If the colloidal particles are in a solvent containing a large concentration of ions (high ionic strength) the colloidal particles can approach each other closer than if the ionic strength were low.

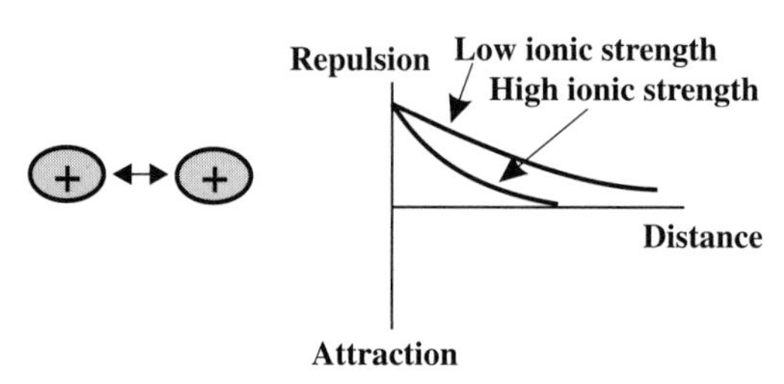

Fig. 6.6 The repulsive forces between two particles.

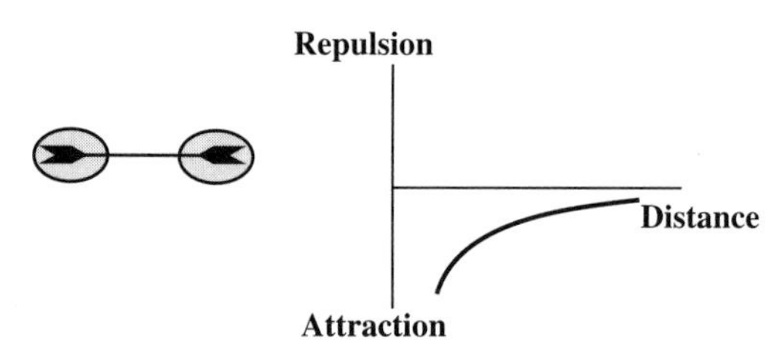

Fig. 6.7 The attractive forces between two particles.

If we add the forces V_A and V_R together we can produce a curve representing the overall forces acting on the particles (Fig. 6.8). If the net repulsion predominates, the particles remain apart. At higher ionic strength the double layer becomes compressed (i.e. the V_R curve drops off more rapidly as it gets further from the particle) and the net particle interaction is one of attraction (Fig. 6.9). The energy barrier stopping the particles coming together has disappeared—the particles aggregate.

Aggregation of colloidal particles can therefore be induced by increasing the ionic strength.

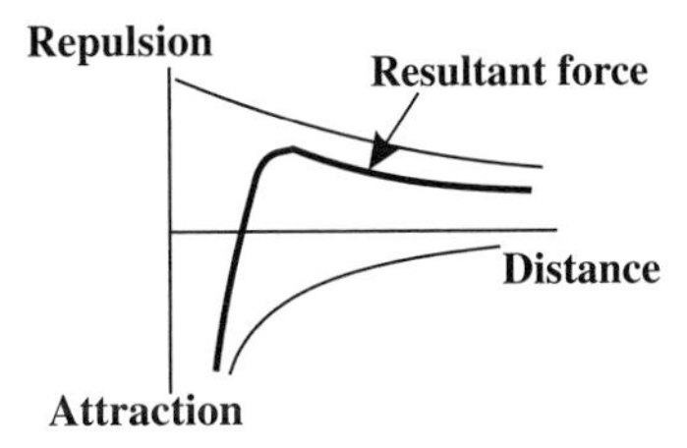

Fig. 6.8 The combined forces acting on two interacting particles in a low ionic strength medium.

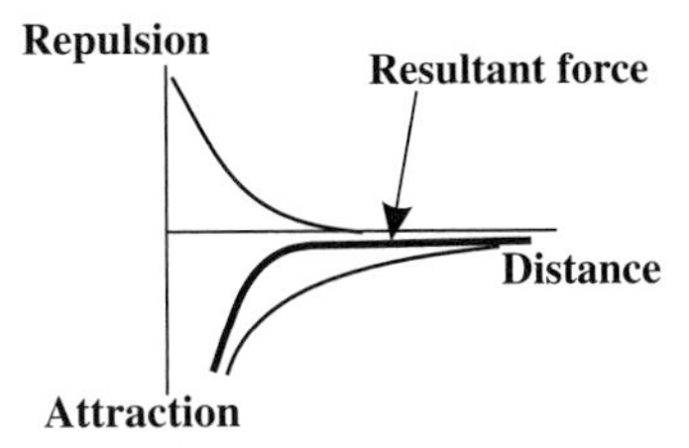

Fig. 6.9 Combined forces acting on two interacting particles under high ionic strength conditions.

Classes of colloid. There are three main classes of colloid:

Hydrophilic colloids are normally very large molecules such as proteins or ions which interact strongly with water.

Hydrophobic colloids interact less strongly with water but are stable because the particles repel each other.

The third class of colloids, that which is taken up by typical soaps (Fig. 6.10), is the **association colloids**. These are collections of ions and molecules which associate as micelles. The stearate ion consists of a hydrophobic carbon chain and a hydrophilic CO_2^- head group. The head groups interact strongly with the water whilst the hydrophobic tails group together away from the water. This results in the formation of micelles (Fig. 6.11).

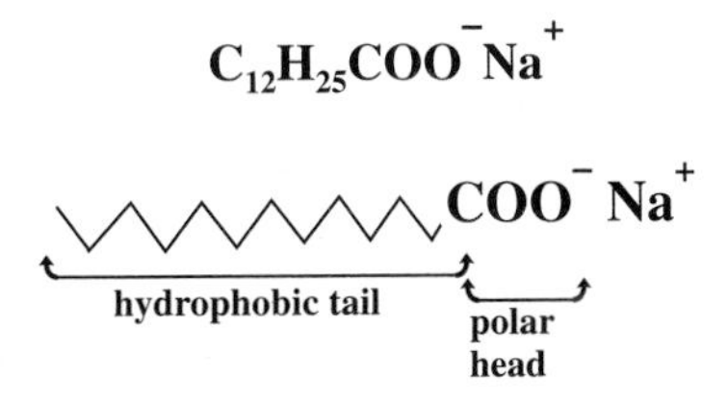

Fig. 6.10 A typical soap, sodium stearate.

The region inside this micelle is hydrophobic and an attractive environment for hydrophobic fatty molecules.

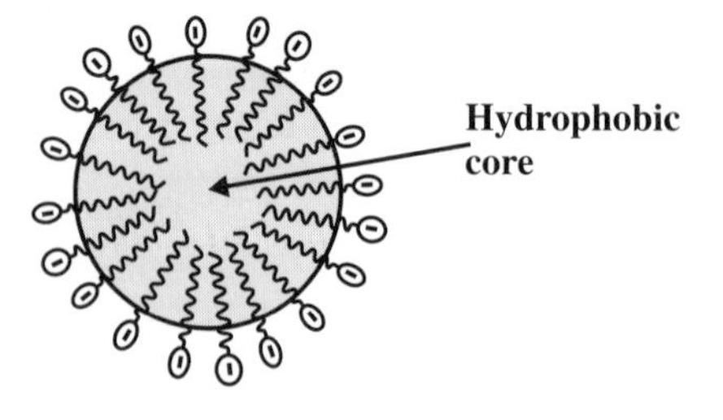

Fig. 6.11 A micelle.

The aggregation of colloids

It is not unusual to find the words coagulation or flocculation used instead of aggregation. It has however been argued that flocculation should only be applied to situations in which the particles aggregate as a result of the intertwining of fibrous particles (cf. in the wool industry).

Whilst colloids are generally stable species, they depend for their stability on surface charges to keep the colloidal particles apart. Any environmental process which neutralizes the surface charge can result in the aggregation of the colloidal particles and cause their precipitation. The most commonly encountered changes which bring about precipitation are those of pH and ionic strength.

The bulk structure of a wet manganese dioxide particle can be thought of as a lattice of manganese and oxygen atoms. At its surface, terminal hydroxide groups are present (Fig. 6.12). Under acid conditions the surface hydroxide groups protonate, resulting in a surface having a net positive charge. Basic conditions on the other hand lead to protons being abstracted from the surface coordinated hydroxide resulting in the manganese dioxide surface having a net negative charge.

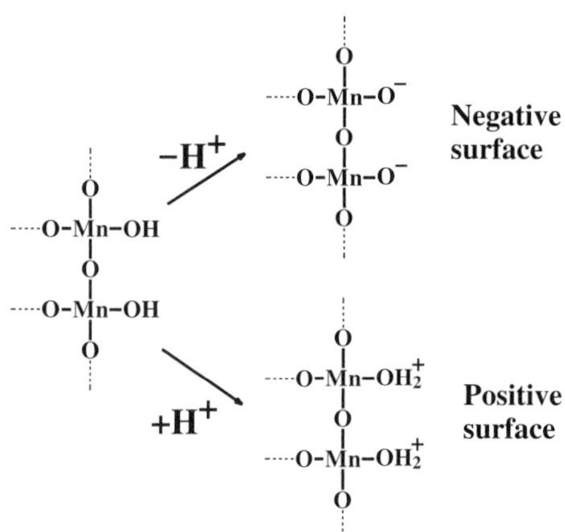

Fig. 6.12 The manganese dioxide surface.

At some intermediate pH there will be a point at which the net surface charge on the particle will be zero, this is called the **zero point of charge**. At this point the charges on the particles will no longer repel each other, the colloidal particles will be able to come together, and precipitation will occur.

The treatment of wastewater

Domestic and industrial demand for clean water is apparently insatiable yet the capacity of natural processes to purify the waste resulting from its use is limited. Waste has therefore to be treated before it is returned to the environment or reused. In many countries the extent to which municipal sewage is purified before release is dictated by standards laid down to protect the quality of the receiving environment. The treatment normally consists of up to three distinct stages: primary, secondary, and tertiary treatments.

Primary treatment

In a typical primary treatment regime insoluble matter is removed first by screening and then by allowing grit to settle out in a settling tank. Further removal of particulate material occurs during **primary sedimentation**. Small particles and colloidal material aggregate and fall to the bottom of a tank whilst grease and oils, waxes and soap materials float to the surface. Both the sedimented and floating materials can then be physically removed.

Secondary treatment

This is generally employed to reduce the amount of organic matter that is still in the water after primary treatment. If released into a river before secondary treatment the receiving water would have to accept a harmful biological oxidation demand.

Biological oxidation demand is a measure of the quantity of organic matter in a sample of water. It is measured by assessing the quantity of oxygen which is used up in its degradation. The BOD test utilizes aerobic microorganisms to degrade the organic matter in a sample over a fixed time period (typically 5 days) at 20 °C. 2 bottles of sample are taken, one is analysed immediately for its dissolved oxygen content, the other is analysed after a 5 day incubation. The difference between these two dissolved oxygen values gives the BOD of the sample.

The process of secondary treatment exploits the ability of microorganisms to oxidize the organic matter. In its simplest form a **trickling filter** is employed in which wastewater is trickled, under aerobic conditions, through a bed of rocks covered with microorganisms (Fig. 6.13).

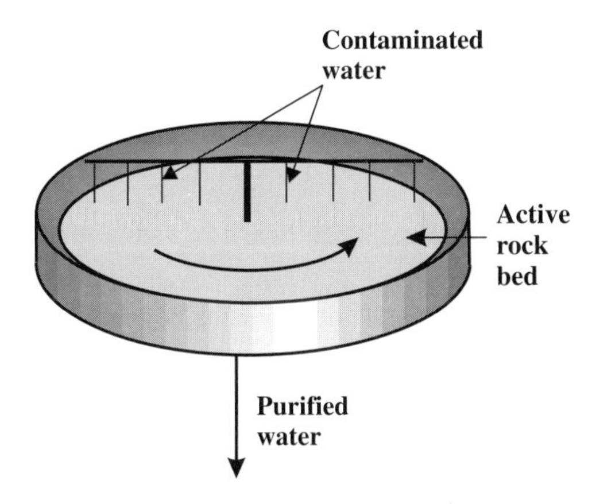

Fig. 6.13 A trickling filter employed in secondary waste treatment

A more versatile approach to achieving the reduction in BOD is the **activated sludge process** in which microorganisms convert the waste organic matter to carbon dioxide and microbial biomass in an aerated tank (Fig. 6.14).

The process results in organo-phosphorus compounds being converted to phosphates and organo-nitrogen compounds to ammonia and nitrate. Organic material in the water ends up as two main products: carbon dioxide, which is vented to the atmosphere and solid biomass or sludge. The sludge itself now

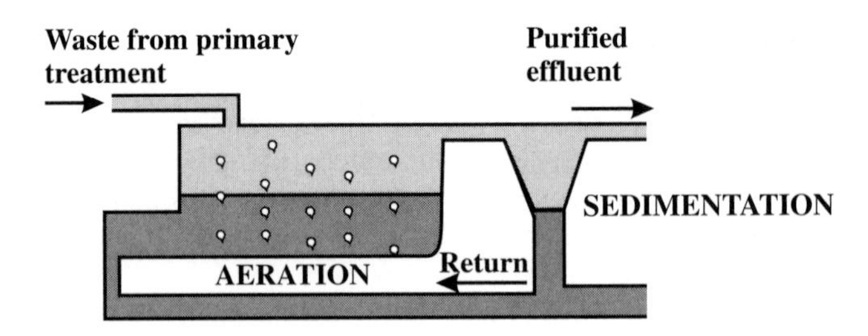

Fig. 6.14 The activated sludge process

The sludge can contain high levels of heavy metals and care must therefore be taken to ensure that this does not give rise to soil contamination.

has to be disposed of. As it is 99% water it is normally first dried and then incinerated, used as land fill or increasingly as a fertilizer.

Removing colloidal material

Whilst sedimentation removes the larger particles, significant quantities of colloidal material remain in the treated sewage. In circumstances when the presence of this remaining colloidal material is unacceptable, chemical measures can be taken to aggregate the colloid so that it can be removed by sedimentation. This is achieved by destabilization of the colloid by the addition of chemical reagents such as aluminium or iron salts.

There are four main ways to destabilize colloidal material:

Compression of the double layer

This is achieved by high concentrations of counter ions having a charge which is the opposite of the colloid primary charge. The effectiveness with which metal ions are able to destabilize negatively charged colloids increases markedly with the charge of the counter ion. Compared with Na^+, Ca^{2+} is 100 times more effective and Al^{3+} 1000 times more effective.

Adsorption and charge neutralization

Dodecylammonium ions ($C_{12}H_{25}NH_3^+$) have a charge of 1 but are much more effective (1000 times) than Na^+ at destabilizing colloids. The dodecylammonium ions adsorb onto the surface, cancelling the effective charge and leaving a new surface which does not interact strongly with the water.

Enmeshment in a precipitate

$Al_2(SO_4)_3 + 6H_2O \rightarrow 2Al(OH)_3 + 6H^+$

Addition of metal salt such as $Al_2(SO_4)_3$ or $FeCl_3$ causes rapid precipitation of the metal hydroxide into which the colloidal particles become enmeshed.

Adsorption and interparticle bridging

This is normally carried out using synthetic organic polymers, such as partially hydrolysed polyacrylamide. Such compounds have chemical groups which can interact with more than one particle causing a linkage between colloidal particles and their flocculation.

The aggregated colloidal material falls to the bottom of the tank leaving the clarified water to flow on to the next stage.

6.5 Case studies

Is seawater saturated with barium sulfate?

Barium sulfate is very insoluble and the amount of barium in the sea could easily be controlled by its solubility. A simple calculation can be carried out to check whether this might be the case. In seawater the concentrations of barium and sulfate are 0.002 and 905 mg dm^{-3} respectively.

If we just take the product of the concentrations ($[Ba^{2+}][SO_4^{2-}]$) we obtain a value of $10^{-8.38}$ mol^2 dm^{-6}; this is close to the solubility product of 10^{-10} and the barium concentration could well be controlled by the solubility of the barium sulfate.

In concentrated salt solutions such as seawater, the effective concentrations (activities) of barium and sulfate would only be about 23 per cent of the measured concentrations due to shielding effects. If this were taken into account in the calculation, the barium concentration would be found to be even closer to its solubility limited level.

Iron in estuaries

As particulate and dissolved material flows from a river through an estuary to the sea, it encounters a rapidly changing chemical and physical environment. The river water is rapidly changed by mixing with seawater. When just 10 per cent seawater is added:

1. The salinity increases from almost zero to 3.5. This increases the ionic strength of the solution causing colloids to aggregate, leads to the formation of chloro-complexes due to the high chloride ion concentration in seawater and causes the desorption of both anionic and cationic species from solid particles.

2. The pH changes, probably from about 6–7 to 8. This will lead to changes in solution speciation, precipitation/dissolution, etc.

There are several aspects of the behaviour of the dissolved iron in estuaries that are rather unexpected. Firstly, the concentration of iron in many rivers greatly exceeds that 'allowed' by the solubility of iron(III) oxides. This is due to the iron in the river being colloidal, allowing the iron to pass through the 0.45 μm cutoff filter that is conventionally used to distinguish between dissolved and particulate material.

Whereas colloidal iron oxides would normally be expected to have a net positive charge, riverine iron oxide colloids are negatively charged. Riverine colloids are not however present in the pure conditions of a physical chemistry experiment. Being positively charged they attract a surface coat of dissolved organic matter. The excess negative charge of the acid groups on the organics confers a net negative charge to the colloid.

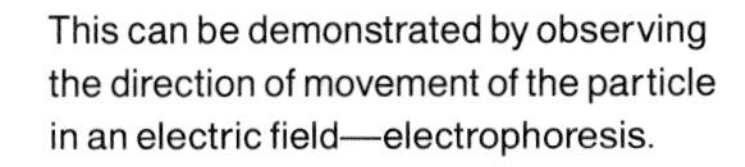

This can be demonstrated by observing the direction of movement of the particle in an electric field—electrophoresis.

When the concentration of 'dissolved' iron has been measured in a number of estuaries, its behaviour has been found to be non-conservative (Fig. 6.15),

Such behaviour is typical of most estuaries including those of the rivers Rhine (Germany) and Beaulieu (UK).

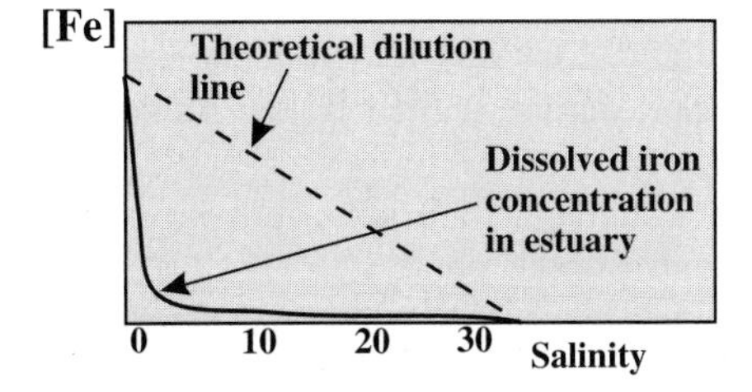

Fig. 6.15 Non-conservative behaviour of 'dissolved' iron in the estuarine environment.

with much of the riverine iron being deposited almost as soon as it encounters the saline water.

As described earlier, destabilization of a colloid requires only a comparatively small change in ionic strength to compress the double layer causing the particle to aggregate and precipitate from the sol.

Sulfur chemistry and the sediments

Sulfur is a widespread element which occurs naturally in a number of chemical forms; as sulfide ores such as pyrite, sulfate ions in seawater and minerals such as gypsum, gaseous oxides of sulfur emitted by volcanoes, sulfur-containing amino acids in proteins and as dimethylsulfoniopropionate (DMSP) in many marine macroalgae and phytoplankton.

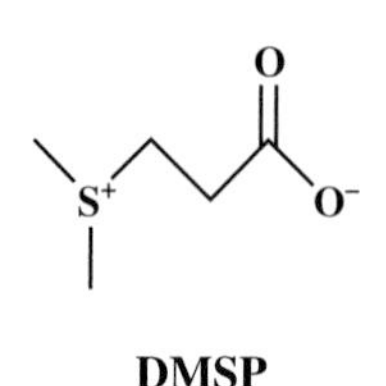

DMSP

Sulfur is a critical element in controlling the fate of trace, and often polluting elements, in the aquatic environment as it can control their solubility. The extremely low solubility of many heavy metal sulfides can ensure that instead of being widely distributed, mobile and readily available to aquatic biota, they are tied up within the bottom sediments.

Sulfur in aqueous solution

The chemical form in which sulfur is present in aqueous solution depends on the acidity and the oxidation potential of the system. In oxidizing conditions there is a tendency for sulfur compounds to oxidize to sulfate (SO_4^{2-}).

As with all chemical systems, whilst under such conditions there may be a strong thermodynamic driving force for the sulfur to be oxidized to sulfate, this process may be slow.

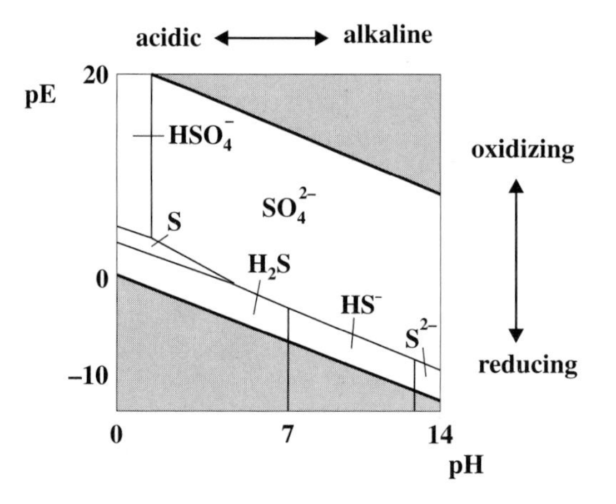

Fig. 6.16 pE/pH diagram for sulfur species

The next common form along the reduction scale is elemental sulfur, which has an oxidation state of 0. Under even more reducing conditions, as found in anoxic (without oxygen) marine sediments, the stable sulfur form is sulfide (oxidation state -II).

In reducing sediments the purely chemical abiotic formation of sulfide from sulfate is extremely slow. The presence of dissolved sulfide is normally only associated with the presence of sulfide minerals and/or the biological reduction of sulfate by sulfate-reducing bacteria.

H_2S dissolves in water and an equilibrium is established between the dissolved ($H_2S(aq)$) and gaseous phase H_2S ($H_2S(g)$) given by:

$$H_2S(g) \leftrightarrow H_2S_{(aq)} \qquad K = 0.99$$

In solution the hydrogen sulfide dissociates giving HS^- and S^{2-}. The proportion ending up as S^{2-}, HS^- or H_2S depends on the pH:

$$H_2S \leftrightarrow HS^- + H^+ \qquad pK_1 = 7.02$$
$$HS^- \leftrightarrow S^{2-} + H^+ \qquad pK_2 = 13.9$$

A significant effect of these equilibria is that there will always be some H_2S present in solution at any pH. This may be insignificant at very high pH but as the pH gets lower the proportion will increase. Below pH 7.02 (25 °C, I = 0) the principal species in solution is H_2S:

Which is why sulfide solutions smell of H_2S.

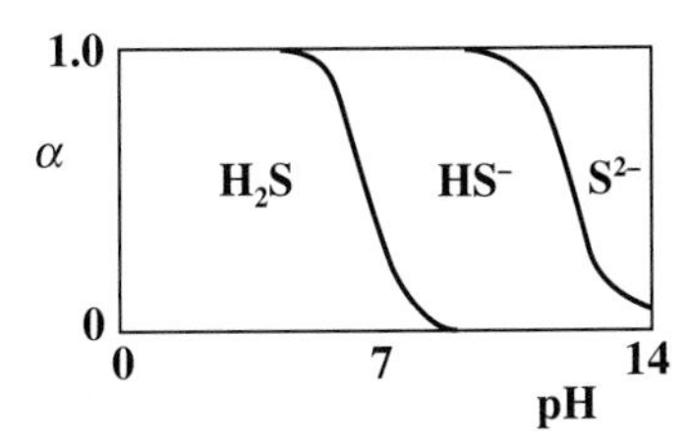

One of the most common fates of dissolved sulfide is for it to react with metal ions in solution to form a sulfide salt.

$$M^{2+} + S^{2-} \leftrightarrow MS$$

Many of these metal salts, and particularly those of metals in the right half of the periodic table, are highly insoluble leading to the retention of metals in reducing sediments as their metal sulfides.

K_{sp} of sulfides at 25 °C

Ag_2S	1.0×10^{-49}
PbS	8.4×10^{-28}
CdS	3.6×10^{-29}
NiS	3.0×10^{-21}

Sulfur in seawater

The concentration of sulfur in the sea can be considered to be an essentially constant proportion of its salinity. Almost all the sulfur in the oceanic water column is present as sulfate, a typical concentration being 2.712 g/kg. The sulfate ion may be partially present in solution as ion pairs such as $NaSO_4^-$, $CaSO_4$ and $MgSO_4$. Sulfide is readily oxidized and the levels to be found in well oxygenated surface waters are extremely low.

On thermodynamic grounds it should not be possible for a well oxygenated water to contain measurable sulfide levels: this must be a non-equilibrium case controlled by the slow kinetics of the oxidation reaction.

A number of marine algae (both macro-algae (seaweed) and phytoplankton) in coastal waters and deep ocean surface waters take in sulfate from the water and convert it to dimethylsulfoniopropionate (DMSP). This compound is believed to be used by the algae to control their internal osmotic pressure and possibly to act as a cryoprotectant. The DMSP is released by the algae breaking down to release dimethylsulfide ($(CH_3)_2S$) into the atmosphere which in turn is eventually oxidized to sulfate.

The release of $(CH_3)_2S$ into the atmosphere can result in acid rain. It has also been suggested that, through the formation of cloud condensation nuclei, it may regulate climate.

Sulfur in sediments

A profile through the water column and into the underlying sediments will generally show well oxygenated water overlying the sediment. The surface of the sediment will therefore be well oxidized and usually a light brown colour. Within a few centimeters of the surface the sediments typically become darker coloured as the rate of oxygen supply fails to match its rate of utilization and the available oxygen levels suddenly drop off to essentially zero—the sediments become anoxic. Anoxic conditions are chemically reducing and, most importantly, the life forms which are best adapted to (and form) such sediments are those which instead of using oxygen find an alternative means of obtaining energy. Such conditions are therefore to be associated with biological activity dominated by methanogenic and sulfate-reducing bacteria. Marine sulfate is therefore reduced to sulfide. These sulfide species can remain in solution, vent into the water column above the sediments or combine with a metal ion such as iron giving a poorly soluble metal sulfide species. A typical marine sediment profile will therefore show the presence of dark bands of poorly soluble metal sulfides below the oxic surface sediments. Once deep enough in the sediments a point must eventually be reached where the underlying sediment bed contains little organic matter and the generation of sulfide essentially ceases.

6.6 Problems

6.1 Calculate both the mass and volume of oxygen that would be required to fully oxidize 1 kg of carbon.
Assuming that a lake contains 8 mg dm^{-3} of dissolved oxygen, what volume of lake water would be completely depleted of dissolved oxygen if sewage containing 1 kg of carbon were to be discharged into the lake?

6.2 Tin mining has been carried out in Cornwall for many years. The tin ore normally occurs together with iron pyrites (FeS_2), arsenopyrites (FeAsS) and the sulphides of a number of metals such as cadmium and zinc. Following the closure of the Wheal Jane tin mine, the pumps were removed and the mine shafts filled with water. This water now flows out of the mine into streams which eventually drain into the estuary of the River Fal. In the context of this environmental disaster:
   (a) Explain the chemistry behind the following:
      (i) The water coming out of the mine is highly acidic
      (ii) The stream and estuary waters contain large quantities of orange-brown sediment
      (iii) The levels of dissolved cadmium in both the mine drainage and the estuary are high.
   (b) What remedial action could be taken to minimize the environmental impact of this discharge?

6.3 A *podzol* soil is typical of the forest soils found in humid temperate regions.
   (i) describe the main characteristics of a podzol soil
   (ii) what chemical processes lead to the distinctive structure of such soils?
   (iii) use the pE–pH diagram below to predict the distribution of manganese in a podzol soil.

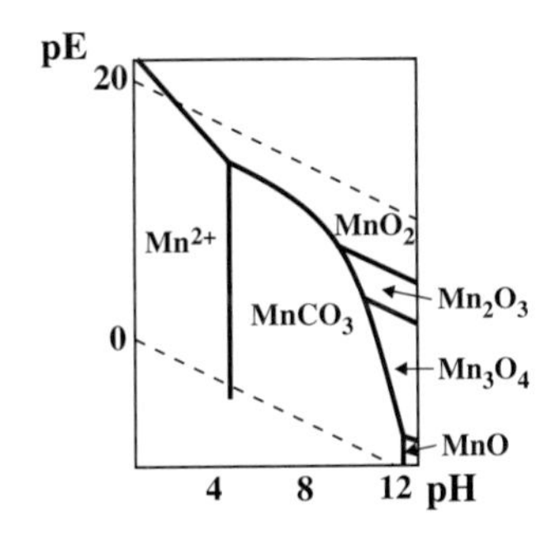

6.7 Further reading

Inorganic geochemistry. Henderson, P. Pergamon Press, Oxford, 1990. ISBN 0-08-020448-1

Environmental chemistry. Manahan, S. E. Lewis Publishers, Chelsea Michigan, 1995. ISBN 1-56670-088-4

Environmental chemistry. O'Neill, P. Chapman and Hall, London, 1993. ISBN 0-412-48490-0

Index